THE BAKED APPLE?
METROPOLITAN NEW YORK IN THE GREENHOUSE

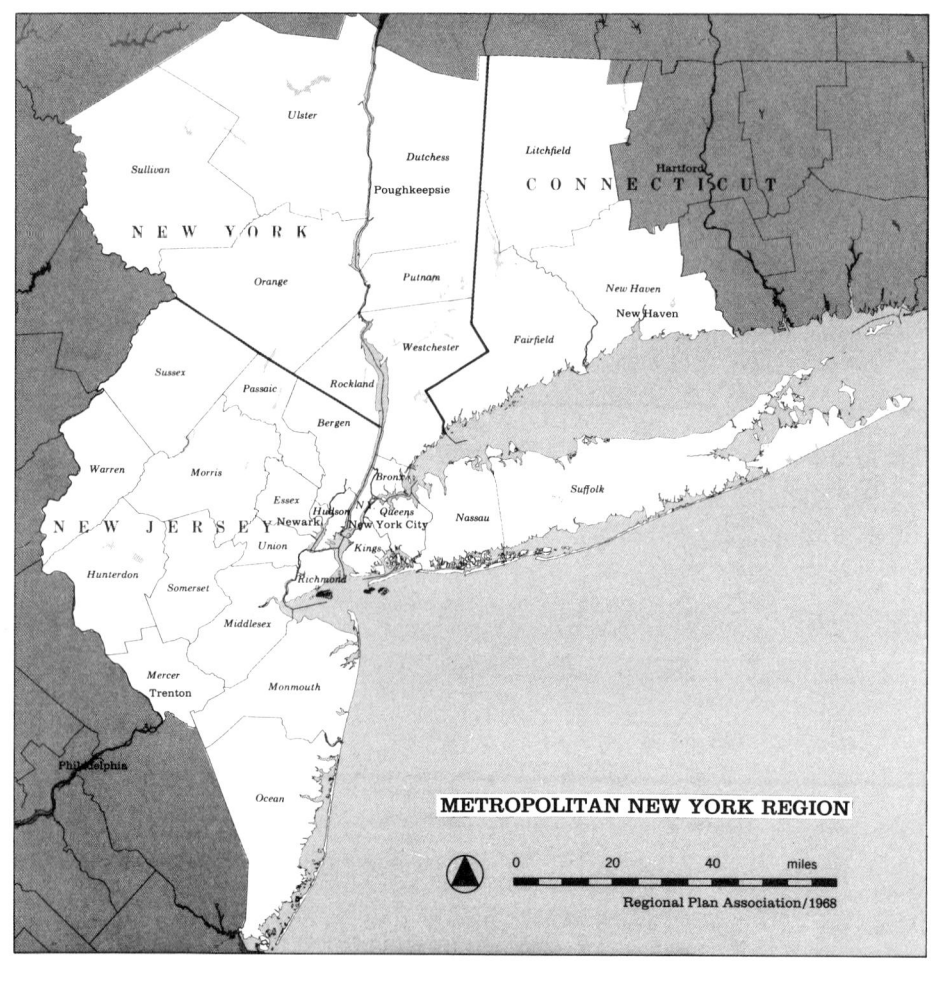

ANNALS OF THE NEW YORK ACADEMY OF SCIENCES
Volume 790

THE BAKED APPLE?
METROPOLITAN NEW YORK IN THE GREENHOUSE

Edited by Douglas Hill

The New York Academy of Sciences
New York, New York
1996

Copyright © 1996 by the New York Academy of Sciences. All rights reserved. Under the provisions of the United States Copyright Act of 1976, individual readers of the Annals are permitted to make fair use of the material in them for teaching and research. Permission is granted to quote from the Annals provided that the customary acknowledgment is made of the source. Material in the Annals may be republished only by permission of the Academy. Address inquiries to the Executive Editor at the New York Academy of Sciences.

Copying fees: For each copy of an article made beyond the free copying permitted under Section 107 or 108 of the 1976 Copyright Act, a fee should be paid through the Copyright Clearance Center, Inc., 222 Rosewood Drive, Danvers, MA 01923. For articles of more than 3 pages, the copying fee is $1.75.

∞ The paper used in this publication meets the minimum requirements of American National Standard for Information Sciences—Permanence of Paper for Printed Library Materials, ANSI Z39.48-1984.

COVER ART: The art on the cover is a detail from a NOAA Landsat satellite photograph and is dated 4 March 1984.

Library of Congress Cataloging-in-Publication Data

The baked apple?: Metropolitan New York in the Greenhouse/editor Douglas Hill.
    p. cm. —(Annals of the New York Academy of Sciences, ISSN 0077-8923; v. 790).
    Includes bibliographical references and index.
    ISBN 0-89766-969-X (cloth: alk. paper). —ISBN 0-89766-970-3 (paper: alk. paper).
    1. Greenhouse effect, Atmospheric—New York Metropolitan Area—Congresses. 2. Environmental policy—New York Metropolitan Area—Congresses. 3. Climatic changes—New York Metropolitan Area—Congresses. I. Hill, Douglas, 1925– . II. Series.
Q11.N5 vol. 790
[QC912.3]
500 s—dc20
[363.73'87'097471]   96-15558
                                                                     CIP

BiC/PCP
Printed in the United States of America
ISBN 0-89766-969-x (cloth)
ISBN 0-89766-970-3 (paper)
ISSN 0077-8923

ANNALS OF THE NEW YORK ACADEMY OF SCIENCES
Volume 790
June 17, 1996

THE BAKED APPLE?

METROPOLITAN NEW YORK IN THE GREENHOUSE[a]

Editor
DOUGLAS HILL

Sponsors
AMERICAN SOCIETY OF CIVIL ENGINEERS, METROPOLITAN SECTION,
INFRASTRUCTURE POLICY GROUP
THE NEW YORK ACADEMY OF SCIENCES, ENGINEERING SECTION
NEW YORK SEA GRANT INSTITUTE
NEW YORK STATE BAR ASSOCIATION, ENVIRONMENTAL LAW SECTION
THE PORT AUTHORITY OF NEW YORK AND NEW JERSEY
REGIONAL PLAN ASSOCIATION
RUTGERS UNIVERSITY, INSTITUTE FOR MARINE AND COASTAL SCIENCES
THE STATE UNIVERSITY OF NEW YORK AT STONY BROOK,
MARINE SCIENCES RESEARCH CENTER

CONTENTS

Welcome. *By* R. W. ROPER	ix
Introduction. *By* D. HILL	xi
Acknowledgments. *By* D. HILL	xiii
Summary, Conclusions, and Recommendations	1

Part I. Technical Presentations

The Greenhouse Effect: The Science Base. *By* A. J. BROCCOLI	19
The Region's Long-Term Economic and Demographic Outlook. *By* R. B. ARMSTRONG	29

[a] Conference held at The Port Authority of New York and New Jersey, New York, New York, 3–4 November 1994.

Waterfront Planning and Global Warming. *By* H. B. HAFF............. 43

Global Warming, Infrastructure and Land Use in the Metropolitan New York Area: Prevention and Response. *By* R. ZIMMERMAN 57

Impact of Global Warming on Water Resources: Implications for New York City and the New York Metropolitan Region. *By* R. ALPERN .. 85

Metropolitan New York in the Greenhouse: Air Quality and Health Effects. *By* L. I. KLEINMAN and F. W. LIPFERT 91

Traffic and Transportation Planning: Strategies for Reducing Greenhouse Gas Emissions in the New York Metropolitan Area. *By* J. C. FALCOCCHIO 111

"Growing" Energy-Efficient Physical Plants in the Greenhouse Era. *By* L. AUDIN... 125

Energy Demand and Supply in Metropolitan New York with Global Climate Change. *By* S. C. MORRIS III, G. A. GOLDSTEIN, A. SANGHI, and D. HILL 139

Part II. Policy Perspective

Emission Reduction in New York State. *By* EDWARD O. SULLIVAN 151

Remarks by Manhattan Borough President. *By* RUTH W. MESSINGER 157

Part III. Scenario Planning

Introduction to Scenario Planning Results. *By* M. J. BOWMAN.......... 163

No Climate Change: "Little Green Apples." *By* R. L. SWANSON, S. REAVEN, A. MOONEY, and S. BOWMAN..................... 177

Medium Climate Change: "Apple Fritters." *By* W. TUSA, P. A. CHIN, and S. TANIKAWA-OGLESBY.................................. 183

Accelerated Climate Change: "Apple Crisp." *By* S. C. MORRIS III and M. H. GARRELL 193

Comments by Review Panel. *By* A. BRYSON, J. FOX, P. JESSUP, R. MEYNINGER, E. A. PARSON, and R. R. RUGGIERI.............. 201

Glossary.. 211

Subject Index... 215

Index of Contributors 221

The New York Academy of Sciences believes it has a responsibility to provide an open forum for discussion of scientific questions. The positions taken by the participants in the reported conferences are their own and not necessarily those of the Academy. The Academy has no intent to influence legislation by providing such forums.

To Gwendolyn, Kathleen, Blake and Breton
and their contemporaries in the 21st Century

Welcome

RICHARD W. ROPER

Director, Office of Economic and Policy Analysis
The Port Authority of New York and New Jersey
One World Trade Center
New York, New York 10048

The Port Authority is indeed pleased to serve as your host and as a cosponsor for this event. We consider the issues that you will examine over the next two days to be critically important to the region's future. Not long ago, as a matter of fact, we convened a series of policy round tables that focused on economic prospects for the New York metropolitan region, looking to the year 2015. Our intent was to begin a dialogue with policy leaders in the region about challenges and opportunities that loomed before us, and ways that we might address those problems, meet the challenges, and begin to think about ways of maximizing the opportunities. With a slightly different focus, you over the next two days will engage in discussions that will result in the development of a range of alternative futures for the region, ones that reflect different environmental realities and the infrastructure investments required to sustain those realities. Your task, like the one we undertook several months ago, is difficult, but the issues we confronted then and the ones that you will be struggling with today can't be ignored. Critical economic, environmental and—yes—infrastructure concerns, if not addressed now, will surely imperil our future.

With this in mind, the Port Authority is currently surveying planned infrastructure investment by public and selected private sector entities throughout the region, investments over the next five years. This information, which has never before been collected, will enable us to begin examining the extent to which future infrastructure needs are being anticipated, and where gaps exist and how those gaps might be addressed. Parallel to this effort is one we are pursuing to gain a deeper understanding of how technology will affect future infrastructure investment. Just last month, as a matter of fact, we released a study which identified seventeen emerging technologies that are likely to significantly affect the Port Authority's existing core businesses and customers over the next fifteen years. I think it significant to note that of the seventeen technologies of critical importance to us, six were related to issues of environmental quality.

In closing, let me just say that what you do here today is most important, and I think it's exciting, insuring that the public discussion not only addresses the problems of today but has as well challenges of tomorrow on its agenda is a most worthy undertaking. And for that reason, I wish you well in the work before you. You've got a long day today.

Welcome to the World Trade Center.

Introduction

DOUGLAS HILL

*Douglas Hill, P.E., P.C.
15 Anthony Court
Huntington, New York 11743-1327*

People used to say that everybody talks about the weather, but nobody ever does anything about it. And all the while, nobody talked about climate change, but apparently we were doing something about it. As everybody now knows, we have been adding billions of tons of carbon dioxide to the atmosphere every year, making inevitable—in the opinion of most of the world's climate scientists—global climate change. In much of the world, people are now talking about climate change and what can be done about it. Since the late 1980s, the Intergovernmental Panel on Climate Change, consisting of hundreds of scientists in many countries, has grappled with the subject and published several treatises on it. More than 150 countries, including the United States, have now signed the United Nations Framework Convention on Climate Change in which they have pledged to take precautionary measures to anticipate, prevent, or minimize the causes of climate change and mitigate its adverse effects. The U.S. Climate Change Action Plan of October 1993, which was prepared under this Framework Convention, declares itself "a clarion call, not for more bureaucracy, but instead for American ingenuity and creativity." So naturally, this program is in part a response to that call by a part of New York's professional community with what we believe to be a different approach to addressing this global problem at the local level.

In the global climate jargon, this conference is concerned with mitigation and adaptation, mitigation being the reduction of emissions of greenhouse gases, in particular carbon dioxide, and adaptation: making the adjustments required by climate change, in this case, in the context of the metropolitan area evolving towards a better future.

Most conferences can be judged by the quality of the papers that are presented, and by that standard our fine set of speakers assures a worthwhile program. What is different about this conference is that we hope to go beyond that. Much of the conference, as you know, is concerned with "scenario planning." We will ask you to take the information that is presented by the speakers, evaluate it, and use it to construct a set of scenarios describing a plausible, internally consistent evolution of the infrastructure of the metropolitan area over a period of decades into a better future. At the conclusion of today's presentations, you will be organized into a number of workshops, each assigned a different assumption as to the future severity of global climate change in this area. The details will be described to you this afternoon after the papers, but bear in mind as you listen to these presentations that they are the fodder for your further work, starting late today.

We will not try to make a judgment in this conference as to who is right about whether and how severe global climate change may become. We will consider a spectrum of possibilities from no change at all to very severe to develop this set of what-if scenarios. The climax of the conference will be the presentation of the scenarios developed in the workshops to a distinguished review panel tomorrow afternoon after lunch.

One other thing that I would like to say is that I am pleased that we have a number of students here in the audience to participate in developing this set of desirable futures, because this is your world that we are planning.

And so with an eye toward doing something about it, let us start talking about Metropolitan New York in the greenhouse.

Acknowledgments

This conference is intended to take a first step toward preparing Metropolitan New York for the possibility of global climate change during the next several decades. It has engaged the interest and support of many people from several organizations. I wish to express my sincere thanks to the principal contributors.

The idea for the conference originated with the Infrastructure Policy Group, Metropolitan Section, American Society of Civil Engineers. It enjoyed the support of two successive chairmen of that group, Christian Meyer and Sotiris Pagdadis, as well as its members including in particular Mohammad Longi, Albert Machlin, Robert Olmsted, Alex Simon, Robert Schumacher, Wayne Tusa and Bolivar Sarmiento-Zamora.

The Port Authority of New York and New Jersey generously provided its facilities and contributed substantially to meeting the cost. Bernice Malione coordinated the Port Authority's participation over many months. Linda Handel handled the meeting arrangements. Brenda Scott and Laura Toole aided in the planning. Chris Zeppi managed the Authority's sponsorship.

The New York Academy of Sciences supported the conference through its Engineering Section, and I especially note the contributions of Richard H. Tourin, Joel Kirman, James Cohen and Victor Wouk. I wish to thank Bill M. Boland, the Academy's Executive Editor, and Joyce Hitchcock, Associate Editor, for their kind cooperation in publishing these proceedings.

Anne McElroy, Director of the New York Sea Grant Institute and Fred Grassle, Director of the Institute for Marine and Coastal Sciences at Rutgers University, aided in the planning and provided funding. Doug Gatlin of the Climate Institute, Erik Mortensen of Columbia University, and Rae Zimmerman of New York University assisted in the planning. Robert Pirani of the Regional Plan Association served as a session chairman. Neal D. Madden, Chairman of the Environmental Law Section of the New York State Bar Association, aided in publicizing the meeting. In addition to several of the participants, Gary Goldstein of DecisionWare, Inc. and Kathleen Kelly of the New Jersey Office of State Planning provided helpful comments on the summary, conclusions and recommendations.

I especially wish to acknowledge the support of the Marine Sciences Research center at the State University of New York at Stony Brook. MSRC not only provided financial assistance but shared in the development of the conference and contributed the main effort in its preparation. Jerry R. Schubel, the former Dean and Director of MSRC, was instrumental in the early stages and provided the inspiration for "scenario planning." J. Kirk Cochran, his successor as Director, continued the support. Lori Palmer did the graphic design. Gina Gartin handled the registration. Philip A. Chin and several others sent out the fliers and otherwise did the heavy lifting. Most particularly, I wish to thank Malcolm J. Bowman, without whom the conference quite literally would not have taken place.

DOUGLAS HILL

SUMMARY, CONCLUSIONS, AND RECOMMENDATIONS

In November 1994, most of the world's climate scientists believed that global climate change due to emissions of "greenhouse gases," particularly carbon dioxide, was very probable.[a] The United States, together with more than 100 other nations, had ratified the United Nations Framework Convention for Climate Change (FCCC) pledging to "take precautionary measures to anticipate, prevent or minimize the causes of climate change and mitigate its adverse effects."[1] The U.S. government had released a Climate Change Action Plan[2] in compliance with the FCCC to reduce emissions of carbon dioxide to the 1990 level by the year 2000, primarily through voluntary actions by industry.[b] Some 30 cities in North America and Europe had pledged to reduce emissions of carbon dioxide by 20 percent by the year 2005 or 2010.[3]

Nevertheless, there was widespread unawareness of the potential problem, preoccupation with more immediate concerns, apathy, or disbelief. Although a scientist in a New York City research organization had been the first to call national attention to the problem,[4] although New York City had been the venue of a United Nations Municipal Leaders' Summit on Climate Change and the Urban Environment,[5] there were no visible indications that New York City or the metropolitan area was taking any direct actions for the express purpose of mitigating future climate change or planning to adapt to it.

This conference, *Metropolitan New York in the Greenhouse: The Baked Apple?*, was held November 3–4, 1994 in New York City as a step toward preparing the metropolitan area for possible global climate change, considered in the context of planning for a desirable future. Investment decisions being made now will determine the infrastructure that exists, for better or worse, well into the next century. It is only prudent that these decisions begin to take into account the possibility of climate change during that period.

PROJECTED CLIMATE CHANGE

It was not the purpose of this conference to make a judgment as to the nature or severity of possible future climate change. Rather, an informed scientific report of the possible range of local consequences based on the 1990 reports of the Intergovernmental Panel on Climate Change (IPCC) was taken as an input to the proceedings.

Briefly, it is *very probable* that there will be a warming of the earth's surface and lower atmosphere, and a cooling of the upper atmosphere. Global mean precipitation will increase because the warmer air holds more water vapor.

---

[a] The assertions in this summary are taken from the written and oral record of the conference on *Metropolitan New York in the Greenhouse: The Baked Apple?*, as documented in the remainder of this volume. The editor is responsible for the selection and integration of the material in this summary and for the conclusions drawn and recommendations made. Only additional material to fill a few gaps in that record is referenced here.

[b] It now appears unlikely that this goal will be attained. See W. K. Stevens, "Trying to Stem Emissions, U.S. Sees Its Goal Fading," *The New York Times*, November 28, 1995, p. 1.

It is *probable* that the warming will be more rapid over land than over water, and that there will be a global rise in mean sea level primarily due to thermal expansion. The general circulation models used to make climate change projections are not sufficiently precise to make local forecasts; projections for the New York metropolitan area can be based only on projections for the central part of North America. In these mid-latitudes, it is *probable* that there will be an increase in soil moisture in winter and a decrease in summer, which has implications for agriculture and water supply.

With *lesser confidence*, it is projected that there will be more frequent showers and thunderstorms, more frequent and intense tropical storms, and less day-to-day and inter-annual variability in mid-latitude storm tracks.

POSSIBLE CONSEQUENCES TO METROPOLITAN NEW YORK

From these general climate projections, some possible *direct* consequences to metropolitan New York[c] can be surmised:

- Public health would be endangered by more intense hot spells and worsened tropospheric ozone pollution.
- New York City's main water supply could be at risk owing to inter-seasonal changes in precipitation, evapotranspiration, and runoff, compounded by salt water intrusion into fresh water aquifers and an advance of the salt front up the Hudson River.
- Waterfront flooding and coastal erosion would be exacerbated by a higher sea level and more frequent and severe storm surges.
- Demand for air conditioning would raise peak electrical loads in summer, requiring more generating capacity and probably causing more carbon dioxide emissions.
- Restrictions on the use of automobiles to reduce carbon dioxide and other emissions would become imperative.
- Similarly, energy conservation measures in buildings would need to be more stringent and universal.

The direct impacts of global warming may be much more severe in other parts of the world, for example, with rising sea level displacing millions of people. With New York's prosperity highly dependent upon its role in the global economy and with New York's population strongly influenced by immigration from overseas, therefore, it may be the *indirect* effects of global climate change that prove to be most significant to the metropolitan area.

THE PLANNING CONTEXT

For a "place-based" integrated assessment of these impacts, we start by describing our context: the economic and demographic outlook and institutional constraints.

---

[c] Metropolitan New York is defined as the 31 counties in New York, New Jersey and Connecticut centered on New York City. This region functions as an integrated, interdependent economy with a self-contained commutershed linking a single housing and labor market.

SUMMARY, CONCLUSIONS, AND RECOMMENDATIONS

The economic fortunes of the New York region are tied primarily to international commerce. The region is the nation's unparalleled leader in services exports, multinational business presence, and global financial services. The region's long-term economic outlook is essentially defined by its competitiveness in, and the viability of, global capital, production management and information-based markets.

Future employment is likely to be characterized by a widening gap between high-end knowledge-based jobs and low-end personal service jobs. The demand for routine skills in producing goods and services, including many blue collar and white collar occupations, will disappear as these jobs go abroad. Despite the weak regional economy in recent years, the population has continued to grow, primarily owing to immigration from overseas propelled by forces of family reunification and people being pushed out of their homelands by economic and political conditions. The result is a growing mismatch between the type of labor force available and that in demand. The prospect is for increased population diversity and multiculturalism in the region, greater immigrant and minority representation in the labor force, and further disparity in income distribution.

Professionals and technicians are increasingly hired on a contingent basis for short-term or part-time employment. Some can combine these opportunities to create sustainable self-employment. For those providing information-oriented services, computer and communications technology make telecommuting a practical alternative, by some estimates accounting for up to 40 percent of jobs.

Existing institutions that support land use and transportation patterns are conceptually linked, but the decisions are not well integrated because different agencies are responsible for them. New York is a "home rule" state in which cities, towns, and villages are not obligated to provide land development patterns consistent with State transportation plans that could reduce reliance on automobiles. Existing land use planning institutions do not afford full coverage or control over land use decisions, such as right-of-way zoning in New York City. The availability of undeveloped land will probably continue to allocate population and job growth. Mechanisms such as those provided by the Intermodal Surface Transportation Efficiency Act of 1991 (ISTEA) and the Clean Air Act Amendments of 1990 (CAAA) can be used to guide planning, development and reconstruction of infrastructure, but global warming effects are not now built in as criteria for decisions.

HEAT, AIR POLLUTION, AND HUMAN HEALTH

Metropolitan New York is now an area of "severe noncompliance" with standards for tropospheric ozone. Ozone levels have not decreased substantially during the last two decades because of increased population and energy use, especially automobile traffic. Continued growth in metropolitan New York and upwind areas will, in the absence of further controls, cause more emissions of nitrogen oxides (NO_x) and volatile organic compounds (VOC)—the precursors of atmospheric ozone—and therefore worse air quality. As in other cities, health risks will still exist at air quality levels that meet current standards.

Possible Impacts

Consistent with the IPCC projections, average summer air temperature in the metropolitan area is expected to increase an average of 1° to 2°C by 2030 and 2°

to 4°C by 2070. In the city, local temperatures due to the "heat island" effect may be higher. Severe summer hot spells in cities kill. A 2°C increase might cause an estimated 500 additional deaths per year, mostly from heat, but about 10 percent of them from air pollution: particulates and ozone.

Ozone is formed by the reaction of VOC and NO_x with sunlight in the atmosphere. Higher temperatures would speed the reaction. Moreover, higher temperatures could promote evaporation of VOC from gasoline and solvents as well as increase the emissions of natural hydrocarbons from trees.

Mitigation and Adaptation

The goals of public policy should be i) healthy and comfortable inside and outside temperatures for the general population, and ii) reduced health-threatening air pollution.

The populations most at risk from summer hot spells and air pollution are the very old, the very young, and the infirm, particularly asthmatics. Air conditioning is recommended as a protective measure, but not fans which can exacerbate heat stress in extreme heat. At present, almost all air conditioning is powered by electricity. With electric peak loads already occurring in summer, more air conditioning would require more electric generating capacity that would likely produce more carbon dioxide. To provide the entire population with electric air conditioning would further increase the needed electric generating capacity and could require that many of New York City's residential buildings be rewired to take the electric load: clearly an undesirable choice.

To provide indoor air cooling, technological alternatives should be encouraged such as air conditioners driven by natural gas and "district cooling." District cooling could take the form of an efficient central source of chilled water circulated through a building complex or neighborhood, or an expansion of Manhattan's steam lines to power local air conditioning units.

Passive measures to reduce local temperatures are most desirable. Architectural improvements can improve the natural cooling of buildings. Urban air temperatures can be lowered by reducing the amount of solar energy converted to sensible heat. One way to accomplish this is by increasing urban vegetation which provides shading and increases evapotranspiration. Another way is to increase the albedo of building and paving materials to reflect more sunlight.[6]

Ozone formation in the troposphere may be reduced by reducing the amount of VOC or NO_x, depending upon which is limiting the atmospheric reaction. NO_x is the product of combustion of fossil fuels, mostly from automobiles, electric power generation, and heating. Fossil-fueled electric power generators must significantly reduce NO_x emissions by the end of the century to comply with the CAAA. NO_x emissions can also be reduced by reducing the number of automobiles on the road. Measures to reduce emissions of carbon dioxide through improved energy efficiency and conservation will also reduce NO_x emissions.

WATER SUPPLY

The sources of New York City's water supply—the upstate reservoirs, the Brooklyn-Queens aquifer, and the Hudson River (for emergency supply)—are threatened by pollution from current activities and future development. Growth

SUMMARY, CONCLUSIONS, AND RECOMMENDATIONS

in the City's demand for water will be constrained not only by the economic and political costs of new supplies but by the prospective cost of expanding the City's wastewater treatment facilities.

Possible Impacts

The New York metropolitan area draws surface water mainly from the adjoining Hudson and Delaware River Basins and from aquifers in coastal New Jersey and Long Island. Overall, the regional water supply is in deficit. With more precipitation in winter and less in summer, the existing reservoir system would be less able to meet summer water demands. Higher sea level would cause greater saltwater intrusion into aquifers near the coast. The salt front would extend further up the Hudson River, threatening emergency water supply intakes. In the Hudson River Basin, many more communities might exercise their right to tie into the City system if precipitation becomes unreliable or if an up-river shift in the Hudson River salt front knocks out the intakes for river-fed supplies such as those of the city of Poughkeepsie.

Mitigation and Adaptation

Greater reservoir capacity would reduce the possibility of summer water shortfalls, but the last attempt to construct a dam—the proposed Tocks Island Dam on the Delaware River—was stopped by Federal legislation in 1978. Moreover, more wastewater emerging from New York City would require construction of additional sewage treatment plants. As a practical matter, metropolitan New York's best option is water conservation which could reduce the demand by 20 percent or more.

In drought years, New York City has frequently placed limits on the permissible uses of water. In 1994, New York City took a step toward structural water conservation by giving rebates to building owners who installed low-flush toilets, toilet flushing being the largest consumer of water in buildings. Other possibilities include fixing plumbing leaks, adding flow restrictors to faucets and showerheads, replacing once-through cooling systems, turning off equipment and appliances that require cooling when not in use, and fixing old steam heating systems. Recycling "gray" water from sinks and showers into toilet flushing is possible, and it can be enhanced by mixing in rain water, although plumbing must be changed to separate the potable water system. Water leakage in the system is also a significant "hidden" consumer of water.

Water continues to be wasted because it is grossly underpriced and not yet fully metered, so waste is not noticed. Meters may cover several buildings rather than individual users of water and cannot be read remotely. Because of the impact on low-income tenants in water-wasteful buildings, metering has not been popular. City billing is irregular at best, making it a nuisance to monitor both usage and savings, but the billing system is now being overhauled. Cost of plumbing upgrades is relatively high, and some water-conservation equipment, such as low-flow toilets, does not work well on older systems.

FLOODING AND COASTAL EROSION

The population density along the shoreline in the New York region is among the highest in the country. The use of facilities now on waterfronts—however vulnerable—is inevitable because of the investment they represent, the disruption that extensive relocation would entail, and the need for waterfront locations for a number of facilities to function. Owing to industrial decline and the shift to containerization for shipping goods, however, much of the waterfront is at present abandoned and decrepit.

Possible Impacts

Added to the historical rate of increase in this area of 3 mm per year, sea level rise caused by global warming is expected to total 18 to 40 cm (7.5 to 15.5 inches) by 2030 and 43 to 93 cm (17 to 37 inches) by 2070. Because of local tidal characteristics, high tides in New York City may increase even more. Coupled with the possibility of more frequent and intense storms, the area would be subject to more severe flooding and coastal erosion.

The metropolitan area is near the top of the national list of flood insurance claims filed every year; many places in the region are flooded in major storms. A typical Nor'easter drives water down Long Island Sound into New York City through the East River. A Category 3 hurricane (winds up to 130 mph), with a storm surge elevation of 7.3 m (24 feet), would flood most of the streets in downtown Manhattan. The subway system, sections of which are already being pumped out 24 hours per day, is particularly at risk, especially PATH near the Hudson River and the IRT Division in downtown Manhattan. Other waterfront infrastructure, such as airports (particularly La Guardia Airport), roadways, bridge access roads, railroads, electric power plants, oil refineries and tank farms, solid waste transfer stations, sewage treatment plants, and combined sewer outfalls would be affected. Combined sewer outfalls are equipped with tide gates to prevent sea water from backing up the storm drains. Fully submerged, these tide gates would fail to operate. Exposed steel reinforcing in roadways and steel bridge decks would corrode in the presence of salt water. Road beds and surfacing materials deteriorate under flooding conditions. Many landfills at the waterfront, thought to contain illegally dumped hazardous wastes, would be partially submerged causing contaminated leachate to reach waterways. Water and sewer pipeline failures commonly occur in wet soils at freezing temperatures.

Outside the City, low-lying coastal areas in New Jersey and Long Island, including the barrier beaches, would be subject to flooding and coastal erosion.

Mitigation and Adaptation

A significant portion of the City's waterfront will become available for reuse during the next 25 years. A revitalized waterfront, in a region with hundreds of coastal miles, should be able to accommodate a variety of interests while contributing to the mitigation of global warming and adapting to rising sea level. These are some waterfront planning recommendations for consideration:

- Plant the waterfront with native and coastal species of trees and shrubs to improve air quality and provide natural cooling.
- Renaturalize unneeded bulkheaded shorelines, in particular to encourage the formation of coastal wetlands which can occur even with gradually rising sea level, possibly serving to reduce the loss of fishery resources.

SUMMARY, CONCLUSIONS, AND RECOMMENDATIONS

- Design greenways and bikeways to provide a setback or buffer that helps protect uplands from flood waters.
- Incorporate flood hazards into zoning, such as limiting new development to higher ground.
- Have ideas ready for politicians when the public is anxious to do something after the next major storm.
- With survival at stake, erect storm surge barriers at the three narrow entry locations—The Narrows, the Arthur Kill, and Throgs Neck—to protect the high-density population and large public investment in infrastructure in New York City and New Jersey.

Whether to armor and restore coastal areas (in the manner of the U.S. Army Corps of Engineers) or allow nature to take its course (in the manner of the National Park Service) has been the subject of polarized public debate. Retreating from the shore will not come easily, as illustrated by the repeated reconstruction at taxpayer expense of Dune Road in Westhampton Beach, Long Island.

The flood hazard regulations of the Federal Emergency Management Agency illustrate a regulatory approach. This is a subsidized program that provides flood insurance to property owners in flood hazard areas, provided that the owner builds to rigorous construction standards that are designed to withstand storm and flood conditions and protect lives. In this way, developing a hazardous area becomes the responsibility of the individual property owner.

Standard structural approaches to flood retardation are land reclamation, sea walls, dikes, flood gates, and breakwaters. The Thames Tidal Barrier near London is the most famous structure to protect a major city from storm surges, but similar structures exist in Tokyo and the Netherlands. The cost of constructing relatively massive levees for urban areas built to U.S. Army Corps of Engineers standards has been about $1 million per mile, but costs are highly variable according to the nature of the structure and the circumstances, and they do not end with the construction.

A comprehensive data base of the vulnerability of transportation infrastructure in the metropolitan area by virtue of its location near waterfronts was compiled by several transportation agencies under the direction of the U.S. Army Corps of Engineers.[7] Similar data should be prepared to help plan the future of other types of infrastructure.

TRANSPORTATION PLANNING

Although no inventory of emissions of greenhouse gases has apparently been taken for the metropolitan area, the two major sources of carbon dioxide are certainly electric power generation and transportation, primarily the automobile. In a world where action is perceived to be necessary to curtail global climate change, restrictions on the use of automobiles to reduce carbon dioxide emissions would become imperative in New York as well as elsewhere.

The New York metropolitan area already faces a future that includes increasingly stringent constraints on its transportation system brought about by the age of the system, the limited flexibility due to limitations on available space, and the requirements to meet ozone standards. Greenhouse mitigation measures will add to such constraints but may, at the same time, help to push the system to reduce emissions of other air pollutants, including precursors to ozone.

The possibility of improved automobile technology is not discussed in this volume. Given the inventory of automobiles at any time, carbon dioxide emission reductions will depend upon traffic and transportation management.

While fuel efficiency at highway speeds is a proper concern for most of the U.S., it is less relevant to New York City's stop-and-go traffic. Even disregarding the extra fuel used in accelerating from a stop, an automobile traveling at 20 miles per hour uses two-and-a-half times the fuel per mile that it would at 50 mph; at 10 mph, five times as much. (Emissions of VOC and NO_x from automobiles are also highest in stop-and-go traffic.) To reduce emissions of carbon dioxide from automobiles in metropolitan New York, therefore, is much the same problem as reducing traffic congestion.

One way to reduce traffic congestion is to reduce the number of vehicle-miles traveled (VMT) in peak periods. On expressways nearly saturated with traffic, for example, a 3 percent reduction in VMT can result in a 15 to 25 percent reduction in congestion. There are a number of ways to reduce peak-period VMT, in the near term by modifying the behavior of vehicle users; for example:

- Introduce peak period congestion pricing, implemented by electronic toll-taking, which would theoretically be the most cost-effective method because it makes travel during peak periods more expensive. (Regular commuters now get a discount at toll facilities, in effect making travel during peak periods less expensive.)
- Promote telecommuting, presently encouraged by the Clean Air Act Amendments of 1990 as an element of the trip reduction program required of employers of 100 or more employees.
- Limit taxi cruising in the Manhattan central business district south of 59th Street. One half of taxi mileage is spent searching for fares.
- Discourage parking subsidies, for example, by requiring employers to offer the cash equivalent to those who do not commute by car.
- Provide nontraditional transit services in suburban areas, such as bus shuttles to railroad stations and express vans in high-occupancy-vehicle (HOV) commuter corridors.

A second way to reduce traffic congestion in peak periods is by increasing the capacity or the efficiency of the highway and street network:

- Reduce recurring delays by providing additional capacity, improved traffic controls, and better management of curb space.
- Reduce nonrecurring congestion, which accounts for up to 60 percent of all delays on urban expressways, by early removal of accidents and by providing information to drivers.
- Follow the example of New Jersey by introducing a State highway access code to discourage strip development by specifying how land developments can have access to State arterial highways.

Through traffic and transportation planning, the growth in daily VMT can be reduced in a number of other ways. In the near term:

- Improve existing transit services, for example, by "trip chaining" where a fare is charged for a time period, allowing a person to make stops and change vehicles without having to pay multiple fares.
- Improve intermodal connections.
- Continue to develop a network of HOV lanes.

For the longer term, change mobility and land use patterns:

- Encourage automobile trip reduction initiatives beyond the CAAA requirements, by mandating a process that requires local jurisdictional decisions over land use to be coordinated with State transportation objectives.
- Develop bicycle networks and facilities, encouraged under ISTEA legislation, especially for the half of all urban trips that are less than three miles long.
- Plan and build new developments with access by transit, bicycles and by foot, as opposed, for example, to the typical suburban shopping mall essentially accessible only by car.
- Improve the freight rail system, especially by providing access east of the Hudson River.

It is likely that the area will continue to depend upon its existing networks of highways, commuter rail systems and subways without new major, region-shaping transportation facilities. This will leave the dispersed travel demand in suburban areas not easily served by public transportation.

ENERGY EFFICIENCY AND CONSERVATION

Electric power generation is the other major source of carbon dioxide emissions in the New York metropolitan area. Most of this is consumed in buildings for lighting, electric motors, HVAC (heating, ventilation and air conditioning), and operating equipment and appliances. Since almost two-thirds of the energy consumed to produce electricity is ordinarily wasted as heat, electricity savings are a leveraged source of reductions in carbon dioxide emissions.

Codes and regulations mandating energy efficiency improvements have had limited success. Building codes that apply only to new structures do little for New York City where new construction and building turnover are relatively slow. The stock of building changes only 1 or 2 percent per year, which means that it would be at least 50 years before the last building is touched. Moreover, present State codes which were last updated in 1991 do not even match present off-the-shelf technology. No one is now enforcing the New York State energy code.

Strictly for dollar cost savings, however, carbon dioxide emissions from commercial buildings in New York City can now be reduced substantially by state-of-the-art energy efficiency measures. In a typical ten-story, 100,000 sq ft "glass tower" building built in the 1950s or 1960s, many such measures can be introduced that in total would reduce electrical usage by up to 60 percent and fuel usage by about 12 percent at a cost of $1–2 million. These would consist of improved lighting, more efficient electric motors, timers to turn off idle equipment, and more efficient HVAC, including a building-wide energy management system to control the services to individual tenants. The payback period for individual improvements is generally 2 to 6 years.

What then prevents energy efficiency measures from being implemented? The barriers are historical, institutional, financial, educational, or managerial, not technical. For example,

- Energy prices are not high enough to encourage investment in energy savings. Except for the poor, direct energy costs are typically on the order of 1.5 to 5 percent of income or budget. A "carbon tax" on energy, penalizing the consumption of fossil fuels (most on coal which has a high carbon content,

least on natural gas which is comparatively low in carbon) would place a higher value on reducing carbon dioxide emissions.
- Electric rates are designed to recover capital costs, so they are high for kilowatts of peak power draw and low for kilowatt-hours of electrical consumption; big energy users actually receive special rate *reductions*.
- Landlords pass along energy costs in rents. Indeed, commercial landlords can sometimes get a mark-up on electricity sales to tenants. Renters have little or no control over heating systems.
- Facilities management is traditionally "low-tech"; only 1 percent of buildings have a professional energy manager.
- Builders gain nothing by spending money on energy efficiency. Homeowners and co-op dwellers often lack the funds for investing in energy efficiency.
- Cogeneration of electricity and heat by independent power producers (which can convert two-thirds or more of the energy content of the fuel to a useful purpose) is resisted by the franchised utilities which see them as competition.

To promote energy efficiency, economic forces are needed that shorten the payback period. Markets for trading pollution abatement credits, which would lead to emissions reductions where they are least costly, should be encouraged. (The credits from the sale of pollution rights also provide a funding stream to reinvest in conservation.) Financial incentives should be restored, such as utility rebates and State and Federal grants that lower the cost of installing improvements.

To provide the option of replacing oil and coal as well as electricity, utility district heating steam lines in Manhattan should be extended above 96th Street, and high-pressure gas mains above 137th Street. Exemptions for City-owned facilities that still burn coal should be eliminated.

For increased energy efficiency, updated energy codes should apply to all but the smallest renovations. Energy tariffs should be written to encourage energy efficiency, not just cost of service. Energy policies, such as economic development rates, should promote energy efficiency, not lower energy rates.

ENERGY SYSTEM

For energy supply, the metropolitan area is heavily dependent upon oil, although the share of natural gas is growing. Coal continues to be burned to heat some City-owned buildings. For the next few decades at least, most energy will continue to come from fossil fuel. New York State's potential hydropower resources are virtually exhausted, and public rejection rules out nuclear power for the foreseeable future. New York State has very limited potential for additional renewable energy, including solar and wind energy, and will find it more difficult to meet any future restrictions on carbon dioxide emissions than many other parts of the country.

Compared to the rest of the country, the metropolitan area is relatively energy-efficient because of its mass transit systems and high-density housing. Nevertheless, as we have seen, there is room for substantial reductions in the use of electricity in buildings, and energy can be saved by reduced use of automobiles and reduced traffic congestion. In New York State, there is presently an oversupply of electricity-generating capacity. For the next decade or so, growth in electricity demand has been expected to be limited by "demand-side management" that

SUMMARY, CONCLUSIONS, AND RECOMMENDATIONS

reduces energy needs. With the current structure of the electric utility industry, however, reductions in electricity demand cause the utilities' fixed costs to be allocated over fewer kilowatt-hours of electricity. This leads to an increase in electric rates which, in the metropolitan area, are among the highest in the nation. Nationwide, there is a trend toward restructuring the electric utility industry. This is expected to increase competition in energy supply, even across the boundaries of regional power pools.

In severe storms, interruptions in electric power can be expected, especially outside New York City where transmission lines are above ground. Should storms and flooding become more commonplace, more local emergency back-up sources of power will be needed.

Energy System Model Results

In broad terms, the source of anthropogenic emissions of carbon dioxide from metropolitan New York is the energy system: the set of technologies that import, convert, transmit, and finally use energy to provide a service such as transportation, heating, air conditioning, lighting, and powering motors, appliances and industrial equipment. Any future reductions in carbon dioxide emissions will have to come from modifying the energy system.

For the conference, a mathematical model of New York State's energy system, NYMARKAL, was modified to distinguish between upstate and downstate. The downstate region, roughly from Poughkeepsie south, is taken to represent the Metropolitan area. The model makes it possible to identify what parts of the New York State energy system would provide required future reductions in carbon dioxide emissions at least cost.

If total New York State emissions were in the future restricted to the 1988 level, future reductions would come mainly, and about equally, from fuel switching and energy conservation. The fuel switching would be mostly by upstate New York electric power plants that now burn coal. The energy conservation would be measures in addition to those found cost-effective in the absence of the emission restrictions. After about 2013, additional renewable sources of energy would contribute to the reduction, and later a small amount of nuclear power would be introduced. With a more severe restriction of a 10 percent reduction in carbon dioxide emissions from the 1988 level by 2023, the contributions from fuel switching and energy conservation would both increase, and the share of renewables would begin sooner and become a bit larger. To achieve very substantial emissions reductions of 20 percent or more, major reductions in the use of fossil fuel for motor vehicles would be needed, coupled with considerably more nonfossil resources such as nuclear energy or imported hydroelectricity.

These NYMARKAL results are like those found elsewhere among industrialized countries which indicate that the least expensive alternatives for reducing carbon dioxide emissions are energy conservation and switching from coal to fuel with lower carbon content, particularly natural gas.

However, New York State now has the highest energy efficiency among the 50 States, measured by Gross State Product per million Btu,[8] and New York City itself—with its mass transit system, multifamily housing, and large commercial buildings—is quite energy efficient. Moreover, there is comparatively little coal burned in New York State to generate electricity; Con Edison and LILCO use none. All these factors indicate that the cost of reducing carbon dioxide emissions in New York State—and the metropolitan area—will be higher than elsewhere.

This suggests that New York participate in "joint implementation" under the Framework Convention on Climate Change in which two or more parties pool their greenhouse gas emissions so that the least expensive mitigation measures taken by one party are paid for by the others. To sequester carbon by reforestation in the tropics, for example, may be much less expensive than some alternatives to reduce carbon emissions in New York. Both parties would benefit if New York paid the tropical country an amount less than it would cost to make the emission reduction here.[d]

The NYMARKAL model also shows that measures to reduce carbon dioxide emissions will also substantially reduce NO_x. The reverse is not necessarily the case, since "end of pipe" emission controls on sulfur and nitrogen oxide emissions often result in higher carbon emissions.

CONCLUSIONS

Four types of conclusions emerge from these proceedings: robust measures, measures that are sensitive the extent of climate change, potential conflicts, and critical uncertainties.

Robust Measures

A number of local measures that mitigate or adapt to climate change are already being taken for other reasons. A need to prepare for climate change only strengthens the case for them, and the uncertainty in the future degree of climate change does not alter their desirability. These include:

Energy conservation offers many benefits: reduced local air pollution, reduced dependence upon foreign sources of energy, and according to its proponents many opportunities for an attractive return on investment. Nevertheless, it seems to require government initiatives and support to persist. For many, energy cost is not big enough to bother with without additional incentives. The poor, who may spend as much as a quarter of their income on energy, lack the wherewithal to make the initial investment.[9] Energy conservation has received fitful government support since the first oil crises of the 1970s. Until recently, New York State promoted electric energy conservation through the Public Service Commission by requiring the cooperation of the franchised electric utilities for demand-side management (DSM) and other programs. With the latest shift in the political winds, local utilities have begun to drop DSM programs because they raise the already high electric rates.

Water conservtion has been required intermittently during droughts. First steps toward structural improvements have begun with New York City's subsidies for low-flush toilets, motivated primarily to avoid building additional sewage treatment plants. Further structural measures such as metering individual users and changes in the plumbing of buildings will be expensive relative to the price of water and

---

[d] With the dismantling of the State Energy Office in early 1995, New York State has apparently retreated from its plans (New York State Energy Planning Board,[8] p. 36) to conduct a study to identify and evaluate policy options to reduce greenhouse gas emissions in the State.

SUMMARY, CONCLUSIONS, AND RECOMMENDATIONS

will take a long time to complete. Protection of upstate watersheds has recently been a priority of the New York City government. The cost barrier of new sewage treatment plant construction may continue to drive incremental improvements in water conservation.

Reducing traffic congestion must be a continuing concern to New York City to avoid strangulation in mobility, economy, and—for the victims of ozone pollution—quite literally. The methods for reducing traffic are well known, and the CAAA and ISTEA are providing the means. Electronic toll-taking provides promising new technology to introduce congestion pricing. In a city that has for decades been unable to summon the political will to charge tolls on the East River bridges that pour traffic into Manhattan, however, the prospects for major reductions in traffic congestion seem remote.

Flood control is already needed, apparently well beyond the general perception, as evidenced by the number of flood control claims submitted each year. The financial district, a center of the city's economic vitality, is particularly vulnerable. A start toward addressing the problem has been made with the interagency preparation of a data base of vulnerable transportation facilities. It should be extended to include other waterfront infrastructure.

Measures Sensitive to Climate Change

Other than flood control needs already evident, the nature of waterfront improvements will depend upon expected climate change, especially the rise in sea level. With little or no change, waterfront improvements provide the opportunity to exchange a decrepit and abandoned shoreline in New York City for multipurpose industrial, commercial, and transportation use together with a recreational, aesthetic, and natural resource. Local government planning has been attentive to the possibilities. These plans should also take into account the opportunities to use shoreline landscaping for cooling tree cover and buffering storm surge and flood waters.

With "expected" sea level rise of 28 to 66 cm (roughly one to two feet) and the possibility of more frequent and intense hurricanes, the necessity for defending or drawing back from the shoreline will have to be faced. As a minimum, land use planning and building codes will need to address the changed circumstances.

In a worst case scenario, with the extreme conditions possible by 2070, decisions would need to be made to harden or sacrifice waterfront areas to contend with a 1 to 1.2 m (3 to 4 foot) rise in sea level. Major structures, such as tidal barriers to protect valuable New York City infrastructure, would need to be considered.

Potential Conflicts

The possible impacts of climate change and measures to address it suggest emerging or worsening conflicts. One source of conflict is competition for scarce resources. Two resources that would become more scarce are water and cool air.

New York City's primary water supply drawn from the Hudson and Delaware River Basins also supplies parts of New Jersey and Pennsylvania, including Philadelphia. Protecting the upstate New York watershed has already been a source of conflict between the City and upstate communities. With water supply shortfalls, many more of these communities may tie into the New York City water supply system as they have the legal right to. The Hudson River, New York City's emergency supply, is the primary water supply of upstate communities like Poughkeepsie.

Cool air in summer is a comfort to many, a necessity for a productive working environment for some, and a possible life-saver to a few: the sick and infirm. Inside cool air is now provided almost entirely by electric air conditioning. As an economic and practical matter, electric air conditioning may not become available to everyone in the metropolitan area. The technological alternatives—district cooling and chillers powered by natural gas—are unlikely to make up the difference.

The allocation of air conditioning is now essentially determined by the marketplace; those who can afford it have it. As we have seen in U.S. cities in the very recent past, summertime heat kills. If air conditioning becomes more a matter of life or death, the question arises as to whether its allocation should be more a matter of social justice. Do people have a right to healthy air?

This issue emphasizes the importance of applying passive measures to achieve cool air. These include changes in architecture and building placement that foster natural cooling. Citywide, quite dramatic reductions in outside air temperature can be achieved through extensive urban tree planting and the use of reflective surfaces on roofs and pavement. A Federal program to promote these citywide measures, *Cool Communities*, is at this writing no longer funded.

The economic measure of the scarcity of resources is their price. To promote conservation, "full-cost accounting" is proposed to value energy and water more accurately. Energy prices would increase if their external costs—pollution damage, reliance of foreign sources—were internalized. Prices for City water would presumably increase if the full cost of providing it were covered. Similarly, congestion pricing—an extra charge to travelers during peak hours—would reduce the use of roadways. In all these cases, the marketplace would then determine the efficient allocation of the scarce resource.

In the context of the future development of metropolitan New York, however, consequences other than economic efficiency deserve a thought. Unless electricity prices elsewhere also included full-cost accounting, for example, the comparative disadvantage of its high electric rates would be exacerbated.

Without global warming, the area faces a future with a large part of its population—those who traditionally performed the routine blue collar and white collar jobs—chronically out of work because of a growing mismatch between job requirements and labor supply. Global warming could well swell this unemployment with an influx of unskilled immigrants escaping the effects of global warming elsewhere. The social implications a large unemployed population enduring hotter summers with less cooling air and water cannot be ignored.

Critical Uncertainties

As an input to scenario planning at this conference, the technical experts prepared a list of critical uncertainties: events or conditions that will shape the future but the outcome of which is unknown. These are the things we most need to know to plan for the future. Some of them may be learned through additional research, some can be influenced, for some only time will tell. These are the forks in the road to the future or, for planning the future, forks in a decision tree.

SUMMARY, CONCLUSIONS, AND RECOMMENDATIONS

Summarized at a more general level, these are the critical uncertainties that will determine the ability of the metropolitan area to mitigate or adapt to climate change:

Global Climate

- Local consequences of climate change, *e.g.*, rise in sea level, effect on water supply
- Federal mandates to reduce greenhouse gas emissions

Economic Resources

- Comparative advantage internationally
- Comparative advantage nationally (taxes, electricity rates)
- Efficiency of local infrastructure
- Capital availability
- Extent of Federal and State support
- Competing needs (education, public health, crime control, *etc.*)
- Nature and location of jobs
- Structure of the electric power industry (availability of inexpensive and low-carbon energy)

Human Resources

- Suitable work force (technical skills, language capability, productivity)
- Informed and responsible citizenry
- Political will and capability
- Attitudes toward living near the water and relocation away from it
- Public acceptance of price rises (energy, congestion pricing, conservation measures) and changes in the quality of life (mass transit, gray water, *etc.*)

Local Environment

- Maintenance of natural and cultural amenities to attract and maintain work force and business

Improvements or Changes in Technology

- More efficient automobiles
- More efficient use of energy
- Development of sources of energy with no or low carbon content
- Nonelectric means of air conditioning
- New devices that will require electricity

Science

- Health effects of heat, ozone pollution, and control measures (*e.g.*, air conditioning)
- Susceptibility of the population due to increased age, *etc.*

RECOMMENDATIONS

To inspire a proper regional response to climate change, a process should be established to work toward a consensus. This process should provide a continuing dialogue among citizens, experts, interest groups, and policy-makers. Its agenda should be:

- to promote awareness of the prospect of global climate change
- to provide a realistic assessment of the needs for action
- to foster coordinated regional plans and action
- to integrate global warming issues into all infrastructure planning
- to further develop analytical tools for evaluation of mitigation and adaptation measures
- to identify needed research
- to monitor regional progress toward mitigation and adaptation
- to provide an educational program for the public
- to determine a basis for financing

The goals of this process should be to get past incremental thinking and the immediate political agenda, to develop a vivid picture of the future so that policy-makers can believe in the possibility and so that it can be scrutinized, and to identify the things we most need to learn. To overcome the limits of short-sighted business and political processes, the public should be offered education in long-term planning that promotes a level of responsibility comparable, for example, to that which has developed toward recycling.

A host organization should be identified or created to maintain this process. This could be a governmental entity with regional responsibilities, such as the Port Authority of New York and New Jersey, or a private entity such as the Regional Plan Association.

A regional climate change action plan should be developed that integrates planning for land use, transportation, air quality and energy. It should take a holistic approach that identifies the linkages among these concerns, as opposed to responding to specific immediate problems. The plan should begin with agreement on goals. Goals of the plan should include the preservation and maintenance of the economic comparative advantage of the region, and the maintenance of the regional amenities necessary to attract and hold businesses and skilled people. The plan should address the question of how financing this region's response should be shared among local, state, and federal levels. It should capitalize on the powers of local government to determine the kind of human settlements we have while facilitating multijurisdictional approaches to capital investment in infrastructure and adaptive and mitigation measures.

The action plan should identify time-phased mitigation and adaptation measures, some of which may be contingent on further information and subject to change as information develops (for "adaptive management"). It should deal

SUMMARY, CONCLUSIONS, AND RECOMMENDATIONS

explicitly with probabilities and timing: gradually or suddenly is important. "Trip wires" should be identified that trigger timely specific policy actions.

The waterfront part of the action plan should take a broad view that considers rezoning for mixed and transitional uses. Locations for possible new types of waterfront industry, such as high-tech recycling a part of the tremendous waste stream emanating from the area, should be kept open. The data base on transportation infrastructure at risk from flooding should be extended to other waterfront infrastructure. Areas subject to inundation or periodic flooding should be mapped taking into account flooding dynamics as well as height above present sea level. For anticipated sea level rise, decisions specific to the place should be made to protect an area or be prepared to sacrifice it. The possibility of harnessing gradual sea level rise to develop new coastal wetlands should be considered.

The energy part of the action plan should develop quantitative data on mitigation alternatives and their costs. As input to the analysis, energy and emission inventories should be completed. The plan must transcend the interests of individual utilities or other interest groups. It should identify specific prospects for the use of local renewable resources and the long-term supply of natural gas. It should be contingent on the future acceptability of nuclear power and uncertainty in the supply of hydroelectricity.

The integrated plan should be followed by coordinated regional action. Climate change considerations should be among the criteria used in zoning decisions and building codes. Building codes should apply to all but the smallest renovations as well as new construction.

To establish a basis for funding and to guide the preparation of educational material, the experience of the Federal Emergency Management Agency should be utilized, recognizing that FEMA prepares for local disasters while any consequences of climate change would be more general.

Measures to mitigate emissions of carbon dioxide should be taken where they are least costly, possibly taking advantage of the principles of joint implementation where more costly emitters share the cost. Adaptation measures should be guided by the benefits as well as the costs.

REFERENCES

1. United Nations Framework Convention on Climate Change, Article 3, Section 3. May 15, 1992.
2. The Climate Change Action Plan. U.S. Government Printing Office, October 1993.
3. INTERNATIONAL COUNCIL FOR LOCAL ENVIRONMENTAL INITIATIVES. 1995. ICLEI's Cities for Climate Protection Enlists Ninety Local Governments. The Cities for Climate Protection Newsletter, March 1995.
4. JAMES E. HANSEN. The Greenhouse Effect: Projections of Global Climate Change. Statement to the U.S. Senate Subcommittee on Environmental Pollution of the Committee on Environment and Public Works. NASA Goddard Space Flight Center, June 10, 1986.
5. Municipal Leaders' Summit on Climate Change and the Urban Environment, United Nations, New York, USA. January 25–26, 1993.
6. U.S. ENVIRONMENTAL PROTECTION AGENCY. 1992. Cooling Our Communities: A Guidebook on Tree Planting and Light-Colored Surfacing, 22P-2001, January 1992.
7. U.S. ARMY CORPS OF ENGINEERS, FEMA, NATIONAL WEATHER SERVICE, NY/NJ/CT STATE EMERGENCY MANAGEMENT. 1995. Metro New York Hurricane Transportation Study. Interim Technical Report, November 1995.

8. NEW YORK STATE ENERGY PLANNING BOARD. 1994. New York State Energy Plan, Volume I, Summary Report, October 1994, p. 61.
9. VINE, E. L. & I. REYES. 1987. Residential Energy Consumption and Expenditure Patterns of Low-income Households in the United States. Berkeley, CA: Lawrence Berkeley Laboratory.

The Greenhouse Effect
The Science Base

A. J. BROCCOLI

Geophysical Fluid Dynamics Laboratory/NOAA
Princeton University
P.O. Box 308
Princeton, New Jersey 08542

INTRODUCTION

Radiatively active trace gases such as CO_2, methane, nitrous oxide, and chlorofluorocarbons warm the surface-troposphere system by increasing the infrared opacity of the atmosphere. Several decades of research, using models of increasing comprehensiveness, suggest that substantial changes in climate will result from the anthropogenic increase of these gases that is presently underway. Thus the potential also exists for substantial societal impacts in a multitude of areas, including urban infrastructure.

Any scientific examination of potential impacts, whether qualitative or quantitative, should begin with an understanding of the process by which projected changes in climate are estimated. It is unreasonable to expect forecasts of potential impacts to be any more reliable than the forecasts of future climate from which they are derived. While this may be obvious, the tendency for scientific research to respect disciplinary boundaries often results in this issue being neglected.

For this reason, a broad overview of the scientific issues involved in estimating future climate change seems an appropriate way to begin considering the effects of greenhouse warming on infrastructure planning for metropolitan New York. This overview will begin with a brief review of the physics of greenhouse warming. This will be followed by a short introduction to climate models and their role in climate change research. Some of the key elements of projected climate change due to greenhouse gases will be enumerated, with emphasis on estimates of climate change for the New York metropolitan area. Finally, several of the issues that contribute to uncertainties in projecting future climate change will be discussed.

PHYSICS OF THE GREENHOUSE EFFECT

A number of trace constituents of the earth's atmosphere are radiatively active in the thermal infrared portion of the electromagnetic spectrum. The most important of these is CO_2; others include methane, nitrous oxide, and the chlorofluorocarbons CFC-11 and CFC-12. There is evidence that human activities are increasing the atmospheric concentrations of these gases. While the volumetric increase of CO_2 is much larger than those of the other greenhouse gases, the combined radiative effect of the other gases is comparable owing to their infrared absorption properties. Wuebbles and Edmonds[1] provide a review of the sources (both natural and man-made), sinks and radiative effects of the greenhouse gases.

Greenhouse gas increases produce a warming of the troposphere by the following mechanism. If one assumes that the earth is in thermal equilibrium, then the incoming solar radiation is balanced by outgoing longwave radiation of equal magnitude. The Stefan-Boltzmann law allows the calculation of an effective radiating temperature (T_e) for the earth that depends only on the amount of incoming radiation and the planetary albedo. The T_e of ~255 K computed by using realistic values for these quantities is lower than the earth's mean surface temperature of ~288 K. Since the tropospheric temperature decreases with height, this implies that the effective radiating level lies well above (5–6 km) the surface. Its elevation above the surface is due to the presence of greenhouse gases in the atmosphere, which act to shield (by virtue of their infrared opacity) the warmer, lower atmosphere from emitting to space. If greenhouse gas concentrations are increased, the atmosphere becomes more opaque to thermal infrared radiation. This increases the degree of shielding, and raises the altitude of the effective radiating level further still. Thus the longwave emission will take place at a lower temperature, resulting in a decrease in the amount of outgoing radiation. This perturbs the thermal equilibrium, so that a warming of the troposphere results that raises the temperature at all levels, including the effective radiating level, and thus restores thermal equilibrium.[2]

Feedbacks involving other parts of the climate system can amplify this direct warming. For instance, the greenhouse gas-induced increase of temperature results in an increase in saturation vapor pressure and, accordingly, the moisture-holding capacity and water vapor content of the troposphere. Since water vapor is also a greenhouse gas, this leads to a further increase in the infrared opacity of the atmosphere and a further warming by the above mechanism. The existence of this water vapor feedback, which provides an important amplification of climate model sensitivity,[3,4] has been supported by an analysis of observational data.[5] Another important feedback is the snow-ice-albedo feedback, in which the warming results in a decrease of the area of snow and ice cover. Since snow and ice have relatively high surface albedos, more solar radiation is absorbed as a result, leading to a further increase in temperature.[6,7] The snow-ice-albedo feedback is also positive. Other feedbacks may also be important, such as those involving cloud amount and height,[4,8] but there is more uncertainty surrounding both the sign and magnitude of these feedbacks.

CLIMATE MODELS

The primary tools for exploring the effects of increasing greenhouse gases on climate are numerical climate models. These models utilize a set of mathematical expressions for the physical principles that govern the climate system (*e.g.*, Newton's laws of motion). Since the climate system is extremely complicated and includes many interactions and subsystems, climate models cannot simulate the full complexity of nature. Instead, plausible simplifying assumptions must be made in order to render the problem tractable. Similar mathematical models are used in other disciplines to simulate the behavior of complex systems. For example, the gradual improvement over the past 40 years in the accuracy of weather forecasts is largely due to the success of numerical weather prediction models, which are the "closest relatives" of climate models.

Climate models have been used to investigate the impact of greenhouse gases on the earth's climate for more than 25 years. The state of the art has progressed dramatically during this period, from relatively simple one-dimensional models of radiative-convective equilibrium to three-dimensional models of the coupled atmosphere-ocean system. The development of a hierarchy of climate models for the study of climate change due to anthropogenic increases in greenhouse gases has led to a variety of research studies. Most of these can be separated into two categories based on their experimental design. Using terminology taken from the IPCC Scientific Assessment,[9] these are equilibrium response studies and time-dependent response studies. Such experiments will be the primary subjects of this section.

Equilibrium response studies typically consist of two climate model integrations, each with a different prescribed greenhouse gas content. These integrations are continued until the simulated climate of each is in balance with its radiative forcing, then the two simulated climates are compared to obtain the response to the change in greenhouse gases. The early studies of the impact of anthropogenic increases in greenhouse gases were equilibrium response studies.

The first of these studies was performed with a radiative-convective model, which treats the climate system as a one-dimensional column. Results from the earliest radiative-convective model experiments[3] illustrated two of the most fundamental aspects of the climate response to increased CO_2: tropospheric warming and stratospheric cooling. They also identified the role of water vapor feedback in amplifying the climate sensitivity. This and similar experiments also yielded sensitivities of roughly 1.5–2.5 K for the change in surface air temperature resulting from a doubling of CO_2.[10–12]

More recent equilibrium response studies have been performed with three-dimensional atmospheric general circulation models coupled with simple models of the ocean mixed layer.[4,13–17] These tend to give larger values for the sensitivity to doubled CO_2, ranging from 1.9–5.2 K.[18] The larger sensitivities can be attributed to the ability of these models to represent a wider array of feedbacks, many of which are positive, such as the snow-ice-albedo feedback. Detailed intercomparisons involving climate models from different research groups indicate that differences in their depictions of some of these feedbacks are responsible for the wide range of sensitivities.[19,20]

Time-dependent response studies use models of the coupled atmosphere-ocean system to simulate a change in climate in response to gradually increasing greenhouse gas concentrations. Such gradual increases are similar to what is occurring in the real climate system. A number of time-dependent response studies have been performed, primarily using general circulation models of the atmosphere coupled with dynamical ocean models.[21–23] In addition to providing information about the global rate of warming that occurs in response to a given rate of increase in greenhouse gas concentrations, an important result from these studies has been the identification of important regional variations in the warming rate that are related to the mixing of heat into the deep ocean.

KEY RESULTS FROM CLIMATE MODELS

In examining climate model stimulations of the effects of increasing greenhouse gases, a number of substantial changes in climate appear with some consistency in many of these simulations. TABLE 1 lists a number of these projected changes in climate, supplemented by some results from a single recent simulation with a coupled atmosphere-ocean model. Most of these changes occur on relatively large

TABLE 1. Some Selected Climate Responses to Increasing Greenhouse Gases (with references)

Climate Response	Reference
Warming of the surface and lower atmosphere	18
Cooling of the stratosphere	18
Increase in global mean precipitation	18
Decrease in area of snow cover and sea ice	18
Increase of winter soil moisture for Northern Hemisphere high latitude continents	18
Decrease of summer soil moisture for Northern Hemisphere midlatitude continents	18
More rapid warming over land than over sea	24
Very slow warming over Circumpolar and North Atlantic Oceans	24
Rise in global mean sea level	25
Decrease in intensity of North Atlantic Ocean overturning	37

spatial scales (*i.e.*, continental or greater), since current climate models have relatively coarse spatial resolution (owing to their enormous computation requirements) and scientific confidence is higher for patterns with spatial scales substantially larger than the grid resolution.

The coarse spatial resolution of existing climate models is a major impediment to making detailed projections of climate-related impacts for a relatively small area such as metropolitan New York. The highest resolution climate models currently available may have grid boxes that are approximately 250 km in size, or almost three times as large as the state of New Jersey. This prevents climate variations on smaller spatial scales from being explicitly resolved. The difficulty is even greater, however, since the degree of agreement among various model estimates of greehouse gas-induced climate change is very poor at spatial scales as small as a grid box.

Confidence in estimates of regional climate change computed by averaging a number of grid boxes may be somewhat higher, and this is the strategy that was used in the IPCC Scientific Assessment to explore climate change at scales smaller than global. For use in constructing a set of climate change scenarios for metropolitan New York, the IPCC estimates[18] for the region identified as "Central North America" are used for those quantities (*i.e.*, temperature, precipitation) that vary spatially. Regional estimates for 2030 are taken directly from IPCC, adjusted to represent anomalies from 1990 rather than pre-industrial values. Those for 2070 were obtained by scaling the 2030 estimates according to the projected change in global temperature. While the nearest boundary of this region is located some distance to the west, the use of this information is justified by the relatively low confidence in the ability of current climate models to resolve climate change on scales smaller than ~1000–2000 km. Extreme caution is necessary, however, when using this information for climate impact assessment.

At the largest scale, the IPCC[24] estimates that the global average temperature in the year 2030 will range from 0.7°C to 1.5°C higher than present, with a best estimate of 1.1°C. By 2070, the global average warming will range from 1.6°C to 3.5°C, with a best estimate of 2.4°C. These estimates assume a "business-as-usual" scenario in which there are no attempts to control greenhouse gas emissions. Primarily because of the thermal expansion of ocean waters, sea level in the year 2030 is expected to be 8–29 cm higher than today, with a best estimate of 18 cm. By 2070, the sea level rise is expected to range from 21–71 cm, with a best estimate of 44 cm. These sea level projections are based on the previous estimates of global temperature change.[25]

For the local region, an estimate of the temperature increase in winter ranges from 1.5°C to 3°C in the year 2030 and 3°C to 6°C by 2070. Increases in summer temperatures are estimated to be somewhat smaller, amounting to 1°C to 2°C in 2030 and 2°C to 4°C by 2070. For winter precipitation, models suggest an increase of from 0–15% by the year 2030. Summer precipitation is estimated to decrease by 5–10%. The combination of these precipitation changes and an increase in evaporation associated with higher temperatures is expected to produce a decrease in summer soil moisture of 15–20%. Much caution is required with regard to these estimates for two reasons. First, our overall confidence in simulated changes in hydrologic quantities is lower than for temperature; second, the aforementioned changes in precipitation and soil moisture are taken from an analysis of mid-continental climate change and may be altered by proximity to the ocean.

As for storminess, climate models suggest an increase in the frequency of convective precipitation (*i.e.*, showers and thunderstorms) with increasing greenhouse gases. There is also some indication of reduced day-to-day and inter-annual variability in the midlatitude storm tracks. As for tropical storms, there is some evidence from model simulations and theoretical considerations that the frequency and intensity of tropical disturbances may increase.[18] All of these results are very speculative and difficult to quantify, as current models cannot resolve the small spatial scales required for the adequate representation of many of these weather systems.

SOURCES OF UNCERTAINTY

The scenarios presented in the previous section represent the current state of the art of climate modeling, but there are a number of significant uncertainties that have the potential to alter the climate change estimates substantially. Because of its complexity, our present knowledge of the ocean-atmosphere-land-biosphere system is limited in many respects. In particular, the following three issues limit the current capability for forecasting future climate: 1) uncertainties in the sign and magnitude of various climate feedbacks (particularly cloud feedback), 2) questions regarding the role of the oceans in climate change, and 3) the possibility of other climate forcing in addition to increasing greenhouse gases.

Climate feedbacks contribute prominently to the total radiative effect of greenhouse gas increases. Feedbacks can be positive, if they amplify the initial response to a perturbation, or negative, if they damp the initial response to a perturbation. An example of a well-established positive feedback is the effect of snow and ice cover. An initial cooling causes the global area of snow and ice cover to increase; this increase causes a decrease in the amount of absorbed solar radiation because of the high reflectivity of snow and ice. The consequent decrease in solar heating produces a further cooling. Radiative damping is an example of a negative feedback, in which the nonlinear dependence of emitted radiation on temperature reduces the magnitude of an initial thermal perturbation.

Feedbacks introduce considerable uncertainty regarding climate model estimates, because potentially important ones may not be incorporated in present models. Furthermore, for feedbacks that are included, there can be considerable variation in their magnitude (and even sign) among different climate models. Since it can be extremely difficult to diagnose the sign and magnitude of feedbacks in the real climate system, these inter-model variations can be difficult to resolve. A specific source of uncertainty concerns the magnitude and sign of cloud feedback, which varies considerably among climate models.[19] Since cloud processes occur on spatial scales much smaller than a climate model grid box, it may be particularly difficult to know which climate model (if any) accurately represents their interaction with the climate system.

The role of the oceans in influencing the course of future climate change is another source of uncertainty. The use of coupled atmosphere-ocean models for climate change studies remains in its early stages, so some of the more interesting results have yet to be duplicated by other climate modeling groups. Similarly, more research is needed to carefully dissect the current results and evaluate the mechanisms by which the oceans influence climate change. Comparisons of model simulations with paleoceanographic data may be useful in this regard. In addition, from a geochemical viewpoint, the oceans are vast reservoirs of CO_2, introducing the possibility of additional interactions not included in present simulations.

Finally, projections of future climate change require a knowledge of future variations, if any, in sources of climate forcing in addition to greenhouse gases. A variety of such sources may be important on a 50–100 year time horizon, including stratospheric aerosols of volcanic origin, variations in solar output, and anthropogenic sulfate aerosols.

Large volcanic eruptions can inject sulfur dioxide gas into the stratosphere, forming sulfuric acid aerosols that can scatter incoming solar radiation, increase planetary albedo, and thus have a significant impact on the earth's radiation balance. The combination of a reduction of total solar radiation with the warming due to the thermal infrared effects of the aerosols results in a small net cooling effect at the earth's surface, which can persist for a few years following an eruption.[26] The inclusion of stratospheric aerosols in climate model simulations has yielded similar results.[27] There is little prospect for incorporating the effects of future eruptions into projections of future climate given the absence of predictive ability for these relatively rare events. Fortunately, the climatic effects are relatively short-lived, so only an unusual coincidence of many eruptions over a period of a decade or more would be expected to have a significant influence on inter-decadal climate trends.

Little is known about possible changes in solar output on time scales of 20 to 100 years, since precise monitoring of solar irradiance by satellites has existed for little more than a decade. Satellite evidence indicates that irradiance varies in conjunction with the 11-year cycle of solar magnetic activity (*i.e.*, the sunspot cycle), with a total variation of approximately 0.1 percent.[28] Proxies of solar magnetic activity, such as sunspot numbers, carbon-14 and beryllium-10, indicate substantial variations on time scales of decades to centuries.[29] There has been some speculation that irradiance changes may be associated with these variations,[30,31] but observational evidence and physical mechanisms have been lacking. However, some evidence has been offered for irradiance variations of up to 0.5 percent on these time scales based on solar modeling and the behavior of other stars similar to the sun.[32] But at present the primary difficulty in documenting these relationships, or past changes in solar irradiance in general, is the short period of precise observations.

Anthropogenic sulfur emissions may also have potential climatic effects through two mechanisms. A direct effect is the scattering and absorption of incoming solar radiation by sulfate aerosols. An indirect effect is the potential of aerosol particles to increase cloud albedo by acting as cloud condensation nuclei. While both of these effects are extremely difficult to assess quantitatively, recent work suggests a net cooling effect at the surface with a magnitude that may be comparable to the change in greenhouse gas forcing over the period of instrumental records.[33-35] While estimates of future sulfur emissions may be available for use in chemistry/transport models to estimate the future variations in aerosol concentrations, the large uncertainties in the magnitude of both their direct and indirect effects makes a quantitative projection of the time variation of anthropogenic sulfate aerosol forcing quite difficult.

One final source of uncertainty is the internal variability inherent to the climate system, which may produce variations in climate comparable in magnitude to the forced variations and further compound the problem. Substantial variability appears to occur in the climate system even in the absence of external forcing, spanning a wide range of time scales. Much of this variability involves interactions between the atmospheric and oceanic components of the climate system. The well-known El Nino-Southern Oscillation phenomenon is an example, with climatic effects in widespread areas around the globe. More recently, models of the coupled ocean-atmosphere system have displayed variability of oceanic overturning (with accompanying atmospheric variations) on inter-decadal time scales.[36] Only when the combined effects of the net change from all the climate forcing mechanisms become large enough to produce a response that is larger than the amplitude of this natural climate variability can anthropogenic climate change be clearly detected and, perhaps, quantified. Based on the present level of understanding, this could happen over the next several decades as the increase in greenhouse gases continues nearly unabated.

REFERENCES

1. WUEBBLES, D. J. & J. EDMONDS. 1991. Primer on Greenhouse Gases. Chelsea, MI: Lewis Publishers.
2. MANABE, S. 1983. Carbon dioxide and climatic change. *In* Theory of Climate. B. Saltzman, Ed. New York: Academic Press.
3. MANABE, S. & R. T. WETHERALD. 1967. Thermal equilibrium of the atmosphere with a given distribution of relative humidity. J. Atmos. Sci. **24**: 241-259.
4. HANSEN, J., A. LACIS, D. RIND, G. RUSSELL, P. STONE, I. FUNG. R. RUEDY & J.LERNER. 1984. Climate sensitivity: Analysis of feedback mechanisms. *In* Climate Processes and Climate Sensitivity. J. Hansen & T. Takahashi, Eds. Washington, DC: American Geophysical Union.
5. RAVAL, A. & V. RAMANATHAN. 1989. Observational determination of the greenhouse effect. Nature **342**: 758-761.
6. BUDYKO, M. I. 1968. The effect of solar radiation variations on the climate of the earth. Tellus **21**: 611-619.
7. SELLERS, W. D. 1969. A climate model based on the energy balance of the earth-atmosphere system. J. Appl. Meteor. **8**: 392-400.
8. WETHERALD, R. T. & S. MANABE. 1988. Cloud feedback processes in a general circulation model. J. Atmos. Sci. **45**: 1397-1415.
9. CUBASCH, U. & R. D. CESS. 1990. Processes and modeling. *In* Climate Change: The IPCC Scientific Assessment. J. T. Houghton, G. J. Jenkins & J. J. Ephraums, Eds. Cambridge: Cambridge University Press.

10. MANABE, S. 1971. Estimates of future changes of climate due to increase of carbon dioxide concentration in the air. In Man's Impact on the Climate. W. H. Matthews, W. W. Kellogg & G. D. V. Robinson, Eds. Cambridge, MA: MIT Press.
11. WANG, W.-C., Y. L. YUNG, A. A. LACIS, T. NO & J. E. HANSEN. 1976. Greenhouse effect due to man-made perturbation of trace gases. Science **194:** 685–690.
12. AUGUSTSSON, T. & V. RAMANATHAN. 1977. A radiative-convective model study of the CO_2 climate problem. J. Atmos. Sci. **34:** 448–451.
13. MANABE, S. & R. J. STOUFFER. 1979. A CO_2-climate sensitivity study with a mathematical model of the global climate. Nature **282:** 491–493.
14. MANABE, S. & R. J. STOUFFER. 1980. Sensitivity of a global climate model to an increase of CO_2 concentration in the atmosphere. J. Geophys. Res. **85:** 5529–5554.
15. WASHINGTON, W. M. & G. A. MEEHL. 1984. Seasonal cycle experiments on the climate sensitivity due to a doubling of CO_2 with an atmospheric GCM coupled to a simple mixed layer ocean model. J. Geophys. Res. **89:** 9475–9503.
16. WILSON, C. A. & J. F. B. MITCHELL. 1987. Simulated climate and CO_2 induced climate change over western Europe. Clim. Change **10:** 11–42.
17. SCHLESINGER, M. E. & Z. C. ZHAO. 1989. Seasonal climatic change introduced by doubled CO_2 as simulated by the OSU atmospheric GCM/mixed-layer ocean model. J. Clim. **2:** 429–495.
18. MITCHELL, J. F. B., S. MANABE, V. MELESHKO & T. TOKIOKA. 1990. Equilibrium climate change and its implications for the future. In Climate Change: The IPCC Scientific Assessment. J. T. Houghton, G. J. Jenkins & J. J. Ephraums, Eds. Cambridge: Cambridge University Press.
19. CESS, R. D., G. L. POTTER, J.-P. BLANCHET, G. J. BOER, S. J. GHAN, J. T. KIEHL, H. LE TREUT, Z. X. LI, X.-Z. LIANG, J. F. B. MITCHELL, J.-J. MORCRETTE, D. A. RANDALL, M. R. RICHIES, E. ROECKNER, U. SCHLESE, A. SLINGO, K. E. TAYLOR, W. M. WASHINGTON, R. T. WETHERALD & I. YAGAI. 1989. Interpretation of cloud-climate feedback as produced by 14 general circulation models. Science **245:** 513–516.
20. CESS, R. D., G. L. POTTER, M.-H. ZHANG, J.-P. BLANCHET, S. CHALITA, R. COLMAN, D. A. DAZLICH, A. D. DELGENIO, V. DYMNIKOV, V. GALIN, D. JERRETT, E. KEUP, A. A. LACIS, H. LE TREUT, X.-Z. LIANG, J.-F. MAHFOUF, B. J. MCAVANEY, V. P. MELESHKO, J. F. B. MITCHELL, J.-J. MORCRETTE, P. M. NORRIS, D. A. RANDALL, L. RIKUS, E. ROECKNER, J.-F. ROYER, U. SCHLESE, D. A. SHEININ, J. M. SLINGO, A. P. SOKOLOV, K. E. TAYLOR, W. M. WASHINGTON, R. T. WETHERALD & I. YAGAI. 1991. Interpretation of snow-climate feedback as produced by 17 general circulation models. Science **253:** 888–892.
21. STOUFFER, R. J., S. MANABE & K. BRYAN. 1989. On the climate change induced by a gradual increase of atmospheric carbon dioxide. Nature **342:** 660–662.
22. WASHINGTON, W. M. & G. A. MEEHL. 1989. Climate sensitivity due to increased CO_2: Experiments with a coupled atmosphere and ocean general circulation model. Clim. Dyn. **4:** 1–38.
23. CUBASCH, U., K. HASSELMANN & E. MAIER-REIMER. 1992. Time-dependent greenhouse warming computations with a coupled ocean-atmosphere model. Clim. Dyn. **8:** 55–69.
24. BRETHERTON, F. P., K. BRYAN & J. D. WOODS. 1990. Time-dependent greenhouse gas-induced climate change. In Climate Change: The IPCC Scientific Assessment. J. T. Houghton, G. J. Jenkins & J. J. Ephraums, Eds. Cambridge: Cambridge University Press.
25. WARRICK, R. & J. OERLEMANS. 1990. Sea level rise. In Climate Change: The IPCC Scientific Assessment. J. T. Houghton, G. J. Jenkins & J. J. Ephraums, Eds. Cambridge: Cambridge University Press.
26. MASS, C. F. & D. A. PORTMAN. 1989. Major volcanic eruptions and climate: A critical evaluation. J. Clim. **2:** 566–593.
27. HANSEN, J., A. LACIS, R. RUEDY & M. SATO. 1992. Potential climatic impact of Mount Pinatubo eruption. Geophys. Res. Lett. **19:** 215–218.
28. WILLSON, R. C. & H. S. HUDSON. 1988. Solar luminosity variations in solar cycle 21. Nature **332:** 810–812.

29. BEER, J., U. SIEGENTHALER, G. BONAMI, R. C. FINKEL, H. OESCHGER, M. SUTER & W. WÖLFLI. 1988. Information on past solar activity and geomagnetism from ^{10}Be in the Camp Century ice core. Nature 331: 675–680.
30. REID, G. C. 1991. Solar total irradiance variations and the global sea surface temperature record. J. Geophys. Res. 96: 2835–2844.
31. FRIIS-CHRISTENSEN, E. & K. LASSEN. 1991. Length of the solar cycle: An indicator of solar activity closely associated with climate. Science 254: 698–700.
32. LEAN, J. 1991. Variations in the Sun's radiative output. Rev. Geophys. 29: 505–535.
33. CHARLSON, R. J., J. LANGNER & H. RODHE. 1990. Sulfate aerosols and climate. Nature 348: 22.
34. CHARLSON, R. J., J. LANGNER, H. RODHE, C. B. LEOVY & S. G. WARREN. 1991. Perturbation of the Northern Hemisphere radiative balance by backscattering from anthropogenic sulfate aerosols. Tellus 43(AB): 152–163.
35. CHARLSON, R. J., S. E. SCHWARTZ, J. M. HALES, R. D. CESS, J. A. COAKLEY, JR., J. E. HANSEN & D. J. HOFMANN. 1992. Climate forcing by anthropogenic aerosols. Science 255: 423–430.
36. DELWORTH, T., S. MANABE & R. J. STOUFFER. 1993. Interdecadal variations of the thermohaline circulation in a coupled ocean-atmosphere model. J. Clim. 6: 1993–2011.
37. MANABE, S. & R. J. STOUFFER. 1994. Multiple-century response of a coupled ocean-atmosphere model to an increase of atmospheric carbon dioxide. J. Clim. 7: 5–23.

The Region's Long-Term Economic and Demographic Outlook

REGINA B. ARMSTRONG

Senior Fellow—Economics
Regional Plan Association
570 Lexington Avenue
New York, New York 10022

INTRODUCTION

An understanding of the likely future consequences of global warming in a particular region, coupled with an identification of the appropriate policy measures for a local response, require a degree of interaction between the physical and social sciences. Physical science tells us that future greenhouse effects will likely be serious in the New York Region by 2050, if not severe. But the extent to which an uncertain range of severity merits our attention now, with new public policies and investment actions, depends upon the probable costs and risks to the future economy and demography of the Region. This chapter begins to investigate these issues by setting forth market-driven expectations of the growth and development of the New York Region to 2020, and then speculating about alternative futures as scenarios that reflect desired conditions, predetermined elements or existing trends, and critical socioeconomic uncertainties.

The geographic area of reference is the 31 County Tri-State Region radiating outward one hundred miles from New York City.[a] The Region has a range of development type and density, and a heritage of natural resources, unmatched in the nation, including seven major estuaries and countless critical habitats of diverse species. As an integrated, interdependent economy enscribed within the jurisdiction of three states and several thousand municipalities, the Region effectively functions as a self-contained commutershed linking a single housing and labor market, with less than one percent of all its work trips destined across the regional boundary.

A long-term forecast of the Region's economy and demography is necessarily based in part upon its historical relationship to the nation, and to known factors pertaining to its competitive advantages or disadvantages with respect to other major regional concentrations of production. Increasingly, however, the New York Region competes directly in the world economy, overriding national constraints and influenced by global forces. Today, the forces of globalization are shaping markets, technology, capital and human resources in unprecedented often unpredictable ways, lending as much uncertainty to predictions of the socioeconomic condition as is global warming to the regional climate.

---

[a] The Region consists of five boroughs of New York City, two suburban counties of Long Island, seven suburban to exurban counties of New York's Hudson Valley, fourteen urban to exurban counties of Northern New Jersey, and three suburban to exurban counties of Connecticut.

THE ECONOMIC PERFORMANCE AND
OUTLOOK OF THE REGION

As the nation's leading gateway to the global economy, the New York Region has vested its economic fortunes in international commerce. Earning 43 percent of the international income of all U.S. businesses, the Region is the nation's unparalleled leader in services exports, multinational business presence and global financial services. In 1990, one in every eighteen wage and salary workers in the Region was employed in a foreign-owned establishment and over one in three Manhattan jobs was directly linked to the global economy. Indeed, the Region's current economic performance and long-term outlook is essentially defined by its competitiveness in, and the viability of, global capital, production management and information-based markets.

The National Context and Global Forces

Following the collapse of property, banking and equity markets in the "overbuilt" 1980s, which affected Europe, Japan and the U.S., a lengthy period of idle capacity, high unemployment and slack demand characterized the deep recessions abroad that have only now begun to show recovery. Throughout the U.S., the resumption of growth in real output (Gross Domestic Product, GDP) that followed the recession of 1990–91 was at first slower and shallower than average, protracting the period of full employment recovery from two years on average to three. Over the past year, however, job growth accelerated in the nation and, coupled with advances in productivity per worker that exceeded historic levels in post-recession recoveries, has produced many more new higher-paying positions in skilled white collar service industries.

The corporate re-engineering process that emerged with full force in the last recession, and continues in the post-recovery, has revamped American industry to compete cost-effectively in global markets. By one estimate, over six million jobs have been slashed from payrolls by downsizing efforts as major goods and service producers, including defense contractors, eliminated workers to trim costs.[b] In productivity terms, the results have meant a six percent increase in average worker output per hour since 1990, a rise now being reflected in real gains in average hourly worker compensation.[c] If corporate re-engineering has displaced labor with more effective production practices, capital and technology, it has also shifted the demand to workers with managerial, technical and information-based skills. Over the last four years, managers and professionals have accounted for three in every four new jobs created in the national economy, compared to less than half in the prior ten years.[d]

In reindustrialization is driven by global competitiveness, and affects both goods and service producers, the high-technology industrial sector of defense contracting has been especially hard hit with multibillion dollar reductions in defense procurement. Since prime defense contractors, and their subcontractors, tend to be spatially concentrated, impacts on regional economies of Long Island,

---

[b] T. J. Spitznas & Associates.

[c] "Statistics Reveal Bulk of New Jobs Pay Over Average," *New York Times*, October 17, 1994.

[d] U.S. Department of Labor, Bureau of Labor Statistics.

southern Connecticut, Massachusetts and Silicon Valley have been particularly severe. Unlike other corporations that have successfully retooled their products and processes, it is unclear as to whether defense-dominated firms will undergo the necessary economic conversion to commercial markets.

As the post-cold war economy unfolds into an era of stiff global competition from newly capitalistic and emerging "open" economies, like China, the ample low-cost labor and available natural resources of developing economies effectively drive down wages of routine production jobs in the U.S. Major American corporations not only seek markets for export abroad, but also off-shore locales for low-cost production, some of which reenters the U.S. like the apparel production of China and the data processing of Ireland. The future of low-skilled routine goods or service production is limited in the national economy because the global forces of factor equalization favor developing areas. Instead, as has recently been demonstrated by the new jobs created, the demand for skills in the U.S. economy will be of a twofold and divergent nature: at the high end, emphasizing cognitive skills, communications, and the ability to manage complex systems, with compensation measured on a knowledge-based scale; and at the low end, stressing in-person services, a pleasant demeanor and a modicum of training, with compensation based on the number of hours worked.

The Region's Recent Economic Performance

Even though the U.S. economy has been moving away from goods production into an information-based, service-oriented economy, and the economy of the New York Region has long specialized in these very activities, the recent growth in employment has not benefitted the New York Region, though national productivity gains have been equalled if not surpassed here. In 1989, the Region entered a period of structural transformation that predated the national recession by six quarters. It then continued to lose jobs for more than nine quarters after the onset of a national recovery. The first sign of a reversal occurred in October 1993, when a four-and-one-half year downward spiral ended in the Region with the loss of 775,000 jobs. Since then, 73,000 jobs have been created in one year, or only one percent of the nation's 5.5 million new employment opportunities generated over the past two-and-one-half years since full recovery.

To a greater degree than elsewhere, the Region has been buffeted by globally induced forces of corporate downsizing, defense conversion, and off-shore placement. Yet more so than elsewhere, the Region is dependent upon the demand for its high-end services being generated in global markets, particularly from the advanced economies of Europe and Japan. Until these world regions have fully recovered from recession, it is doubtful that the Region's economy can equally participate in the benefits of globalization. FIGURE 1 compares the Region's recent job losses to its gains in the preceding decade.

Outlook for Regional Employment Growth to 2020

The New York Region currently provides nearly 8.8 million wage and salary jobs, and more than 1.4 million self-employment opportunities, or collectively 10.2 million jobs and 8 percent of total employment in the national economy. In the peak year of 1989, total employment reached 10.9 million payroll and self-employed jobs or 9 percent of the national economy.

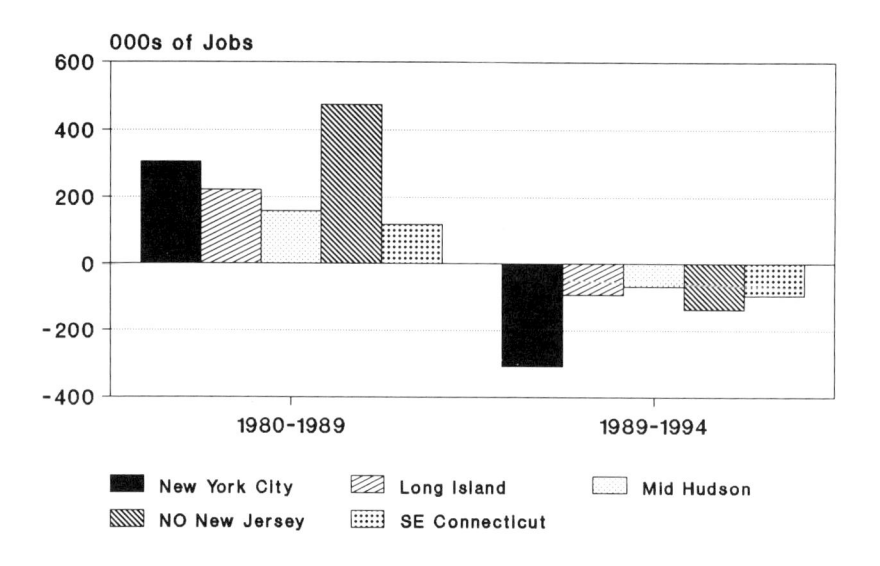

FIGURE 1. Change in payroll employment in the Tri-State Region, 1980–1994.

Given the Region's slow pace of job recovery, the 1989 peak employment level is not expected to again be reached until 2002, or nine years after the turnaround in the economy. If the 1990 decade effectively contributes no new payroll employment growth to the Region, but merely regains previous losses, between 2000 and 2010 payroll employment is projected to rise by .7 percent annually, adding 641,000 new jobs, and between 2010 and 2020, by .6 percent annually or 646,000 new jobs per decade.

Proprietors, or the self-employed, will likely increase more rapidly given evolving new forms of work. Having not declined since 1989, they are projected to add 150,000 new jobs over the 1990 decade, growing by over 1.5 percent per year. Between 2000 and 2010, another 260,000 self-employed jobs are projected at annual increases of 1.6 percent, and between 2010 and 2020, another 350,000 jobs at 1.8 percent per year. Thus, self-employment plays an increasing role in future job formation in the Region, contributing three in every ten jobs generated—or 770,000 of 2.6 million (including full recovery)—between 1994 and 2020. FIGURE 2 depicts the growth in total employment by five year intervals to 2020.

Rapid growth in self-employment, continued expansion of the service industry, and further declines in manufacturing—jobs, not necessarily output—are the significant sectoral trends in the emerging job market. Trade, Finance and Real Estate, and Government will add a modicum of jobs, as will Construction, while employment in Transportation, Communications and Utilities will likely remain stable. FIGURE 3 portrays the relative shifts in composition between present and future employment. Behind these broad classifications, however, lie more striking changes in the nature of work, with implications for skills and earnings, the markets and distributional systems for products, and the spatial organization of the workplace.

ARMSTRONG: ECONOMIC AND DEMOGRAPHIC OUTLOOK

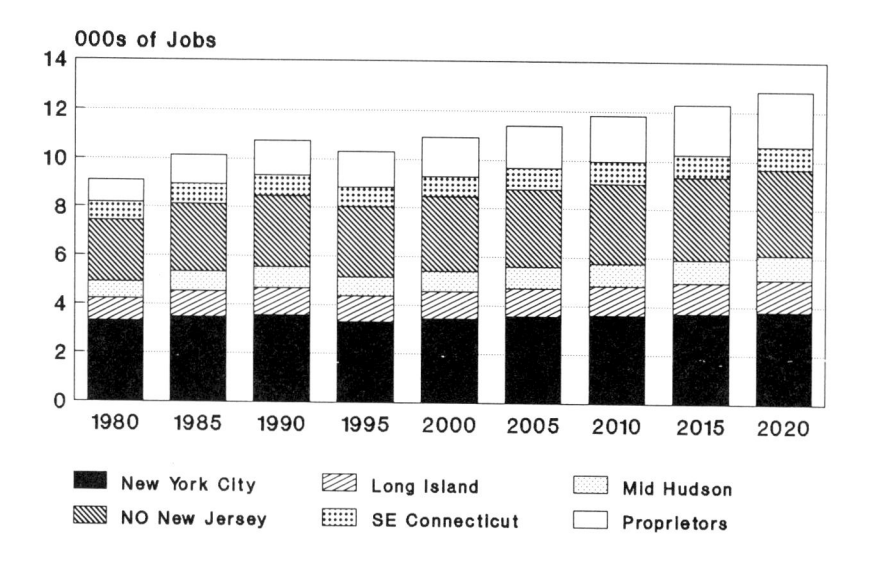

FIGURE 2. Total employment forecast for the Tri-State Region, 1980–2020.

Increasingly, workers are hired on a contingent basis by firms uncertain of their markets, unwilling to pay full benefits, and eager to tap the expertise of available managers, technicians, and professionals by a short term or part-time engagement. The more entrepreneurial or skilled workforce can often parlay these opportunities into sustainable self-employment, but many workers will not find long-term security or adequate labor benefits in the job market. The advent of affordable computer, fax and data retrieval technology for work-at-home environments has meant an increase in telecommuting on a full or part-time basis for

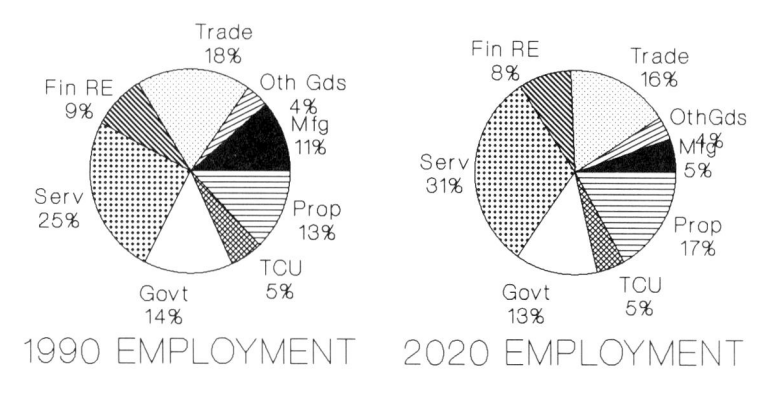

FIGURE 3. Employment composition of the Region, 1990 and 2020.

those in information-oriented services, who prefer to work from home or satellite offices. Some experts think that as many as 40 percent of all occupations may lend themselves to telecommuting on an occasional basis. Conservatively, 5 percent of the Region's workforce will be telecommuting by the year 2000, and the percentage will increase with time. While this presents decided advantages for relieving the transportation network's peak hour congestion, the implications for decentralization on the future functioning of traditional central business districts, like Manhattan, are still unclear.

One school of thought proposes that global markets must be administered from central places: That an essential degree of face-to-face interaction is required in complex and innovative deal-making, and that the physical investment of linking a global network via telecommunication will in fact maintain the central place. Then again, different organizational styles of work characterize different business cultures. As the New York Region evolves into a truly international business center, with European and Asian firms on a par with American businesses, the more socially interactive Asian style of working may affect the more autonomous or cost-conscious European and American approaches.

What is likely to characterize the Region's future growth in employment is a widening gap between the "haves" and the "have nots," propelled in part by the new forms of work. Knowledge-based compensation, technology-driven productivity increases, and competitive advantages of agglomeration will benefit and hold the high-end service producers—including global business managers, financial, legal, design or other business services, professional consultants, the media, arts and cultural community. Many of the self-employed will find themselves able to group and regroup in different competitive configurations like "virtual" corporations. However, the demand for routine goods- and service-producing skills, including many blue collar and standardized white collar occupations, will devolve off-shore, be displaced by technology or drawn on an as-needed basis, or be required to effectively lower wage rates to remain cost-competitive in global markets. Those that provide in-person services to the regional market in retail, health care and social services, government and consumer services, will be in demand locally, but will require relatively less training and capital investment, which will be reflected in their productivity and earnings.[e]

The Distribution of Employment Growth in the Region

Over the 1995–2020 period, the bulk of employment growth in the Region is expected to locate in suburban and exurban areas, although New York City will capture a somewhat greater share in the future than it has in the past. New York City, which lost as many payroll jobs in the 1989–1993 collapse as it created in the 1980–1989 expansion, has historically been declining as a share of the regional job market. Currently, total employment including payroll and proprietary employment amounts to 3.68 million jobs in New York City. Between 1995 and 2020, total employment is projected to grow by 660,000, or one quarter of the Region's total, with roughly half of these jobs locating in Manhattan.

---

[e] Based in part on Robert B. Reich, *The Work of Nations* (Simon & Schuster).

Manhattan

In keeping with changes foreseen in markets and the nature of work, Manhattan will likely retain its edge as a world-city headquarters for foreign and domestic multinational businesses, global capital market functions, world government, the media, cultural and nonprofit institutions. In the future, however, Manhattan may be required to compete with an increased number of world-class centers—not just London and Tokyo—as more diversification occurs in global markets and centers of excellence like Paris and Hong Kong compete in their fields of specialization. Manhattan is also likely to evolve from a producer to a consumer center, strengthening its entertainment, recreational, arts and cultural role to attract more tourists, evening visitors, and regional residents with increased leisure time.

Rest of New York City

The outer boroughs will command an increasing share of New York City's employment growth in the foreseeable future, just as they have weathered the recent losses better than Manhattan. In part, the job attraction to the outer boroughs will reflect the development of subcenters in Downtown Brooklyn, Long Island City, Jamaica Queens, and the Bronx Center. But job growth will also be attributable to the development of enclave economies in the expanding immigrant neighborhoods and the dispersal of personal services, like health care clinics, throughout the boroughs.

Long Island

After a wrenching loss of 93,000 payroll jobs between 1989 and 1994, which represented Long Island's first recession experience, the Island economy is expected to undergo a slow recovery with relatively low growth rates thereafter. By 2002, the 1989 peak of 1.14 million jobs will have been recovered, although the new jobs will be of a different character. Cutbacks in defense procurement nationally meant that Long Island's prime contract volume dropped from $5.3 to $2.1 billion per year, and that major contractors have had to undergo massive downsizing. The largest, Grumman, scaled back its Long Island workforce from 22,000 to 8,000, and will eventually drop to 3,800 employees. The development of a contingent workforce has been especially noticeable on the Island which lends an element of uncertainty to its housing and retail markets. Between 1995 and 2020, Long Island is projected to grow by 415,000 jobs, or 1 percent per year, capturing only 16 percent of the Region's total employment growth. One in every three new jobs created on Long Island will be in self-employment, and three in every five added payroll jobs will be in business, personal and nonprofit services including education and research.

Mid-Hudson

Alone among the subregions, the Mid-Hudson Valley has continued to lose employment in 1994, after dropping nearly one in every thirteen payroll jobs between 1989 and 1993. Corporate re-engineering continues to be the major factor with at least 30,000 jobs lost from downsizing in Westchester County, the major employment center, and additional job losses form IBM in Dutchess and Ulster counties. The long-term future of employment growth in the Mid-Hudson lies in part with the unique amenity advantages the subregion offers to exurban development, second home settlements, and the growth of rural economies focused on environmentally sensitive activity and telecommuting. Between 1995 and 2020, the Mid-Hudson is projected to attract 290,000 new jobs, or 11 percent of the Region's total, which represents an annual growth of one percent. In descending order of importance, major sectors of job growth include: services, proprietorships, and retailing.

New Jersey

The fourteen counties of Northern New Jersey represent the most robust market for new employment growth in the Region. Although New Jersey experienced significant job losses in the 1989–1993 period, dropping 186,000 payroll jobs, it has led the fragile recovery, already regaining 50,000 jobs in 1994. Over the 1980 decade, the New Jersey subregion generated three in every eight new payroll jobs created in the Region. Between 1995 and 2020, the area is expected to provide 1.0 million new payroll and self-employment opportunities or a marginally greater share of the regional total. New Jersey will fare better than other suburban areas of the Region because there is less restructuring in the economy, a greater balance between office and production activities, and its high-technology industries serve export and consumer markets not the defense industry. The strongest New Jersey gains will occur in central areas of Middlesex, Monmouth, Morris and Somerset counties.

Connecticut

Not unlike Long Island, Connecticut has been particularly hard hit by reductions in defense procurement, but it has also paired these losses with corporate relocations from headquarter centers. Connecticut has traditionally specialized in manufacturing, whether as front offices or factory facilities. The continuous loss in manufacturing employment in the Region, likely to still contract by half, bodes poorly for Connecticut. In the future, the emphasis will be on diversification and the attraction of more service activities. Between 1989 and 1994, Connecticut lost 94,000 payroll jobs, or 80 percent of its job growth in the 1980s. By 2020, the subregion is expected to add 215,000 jobs, growing annually at .8 percent, to capture roughly one in every twelve payroll and self-employment jobs created in the Region.

THE REGION'S NEW DEMOGRAPHY AND SETTLEMENT PATTERN

Despite the severe downturn in the Region's economy over the past four years, and the sharp divergence between the regional and national pace of recovery, population growth has continued to occur in New York City and the surrounding suburban and exurban areas. Growth has been propelled by legal foreign immigration to the nation, which has historically shown a strong preference for settlement in the Region. In the past, however, regional growth was driven largely by the demand for labor: As the economy heated up, or cooled down, working-age persons and dependents resettled in or left the area in search of employment. Today, global forces are propelling the new in-migration as a diverse array of foreign-born are pulled into the Region by forces of family reunification and pushed out of their homelands by economic and political conditions. It is not likely that global migration flows will be quelled until sending countries achieve a greater degree of sustainable development.

The Region's Population Growth

In the 1980s, the Region's population grew slowly, adding 650,000 inhabitants to 19.84 million by decade end. Growth was comprised of roughly equal parts of natural increase and foreign immigration, as more than 900,000 nonhispanic whites and several hundred thousand native-born blacks migrated out of the Region to settle elsewhere in the U.S. All told, nearly 1.3 million foreign immigrants were legally admitted to reside in the Region in the 1980s and several hundred thousand returned or resettled outside. Since 1989, as the influx of immigrants to the nation increased, legal admissions to the Region have risen by over 25 percent on an annual average basis. More than 600,000 new immigrants have arrived in the Region during the first half of this decade, an even greater number of residents have out-migrated, and the total population has risen to an estimated 20 million, an increase of several hundred thousand.

Between 1995 and 2020, assuming the continuation of foreign immigration and the levels of employment growth projected, the Region's population can be expected to grow by 2.5 million, or by roughly .5 percent per year to 22.5 million persons. Given the differences in age structure and fertility of new immigrants, the Region may add more than 75,000 persons per year from net natural increase, or the excess of resident births over deaths. But net out-migration of native born residents is likely to continue until past the full recovery, alongside the heavy foreign influx, because the Region's near-term job growth will not be sufficient to adequately absorb all those seeking employment, particularly in higher skill levels.

Compared to the nation as a whole, the Region's population is growing more slowly, though evolving more rapidly in its diversity. Whereas the U.S. population will grow by 23.7 percent under moderate assumptions between 1995 and 2020, population in the Region will expand by 12.5 percent and slip from 8 to 7 percent of the nation's total. As a share of the nation by race and ethnicity, white, black and hispanic groups are expected to decline while Asians will concentrate increasingly in the Region.

Regional Growth Characteristics

As the new demography alters the Region's racial/ethnic composition toward a more heterogeneous society, it also impacts the age structure, household size, labor force composition and settlement pattern. Four mutually exclusive groups define the race and ethnic composition: white nonhispanics, black nonhispanics,

hispanics, Asian and other races. As FIGURE 4 shows, in 1990, two in every three residents were white nonhispanic, nearly equal shares of 15 percent were black nonhispanic and hispanic, and 5 percent were Asian and other races. By 2010, blacks, hispanics and Asians will comprise half of the Region's population, and by 2020, white nonhispanics will represent well under half and hispanics fully one quarter of the population. Asians, the fastest growing minority, are expected to more than triple between 1990 and 2020, while hispanics will nearly double as they become the largest single minority in the Region.

Growth in the Region's population since 1980 has also been characterized by emergence of the "baby boom" generation in prime labor force and house-buying age groups, and movement of the "baby bust" generation into teen and early labor force ages. But the bulge in population growth occurring between 25 and 44 years in the 1980s, as FIGURE 5 shows, also represents the arrival of immigrant population. Between 1995 and 2020, as the "baby boom" matures into middle and late labor force ages, the deep declines of the aging "baby bust" generation will be ameliorated in the Region's prime labor force ages by a continued influx of foreign immigration. Foreign immigrants will also contribute to the growth in youthful and elderly population. On balance, over the next quarter century, the bulk of the Region's growth will occur above 45 years of age, with significant gains foreseen in the retirement ages.

Immigrants will comprise the bulk of growth in the Region's resident labor force between 1995 and 2020. Some immigrants will be adequately prepared to participate in the evolving knowledge-based economy, but others will require remedial assistance to overcome problems of English language and functional illiteracy. In all probability, the Region's mismatch between skills of the labor

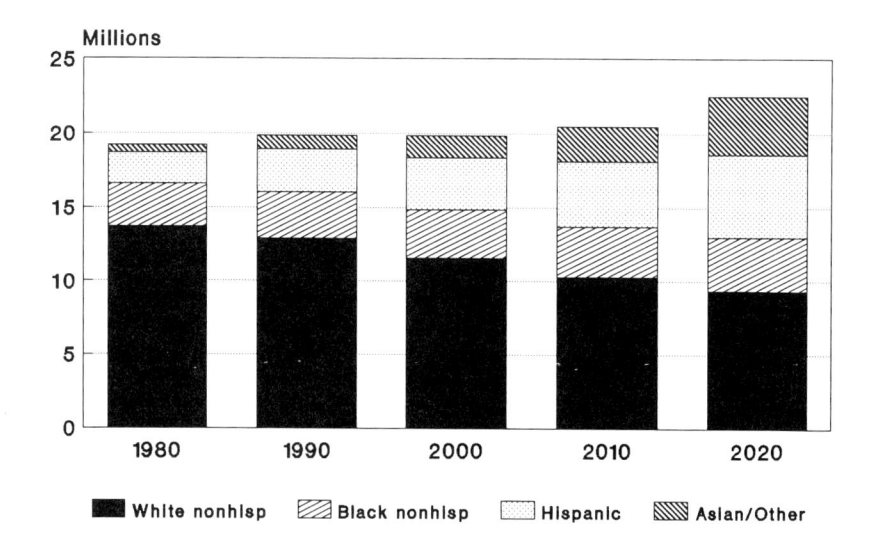

FIGURE 4. Total population in the Region by race and ethnicity, 1980–2020.

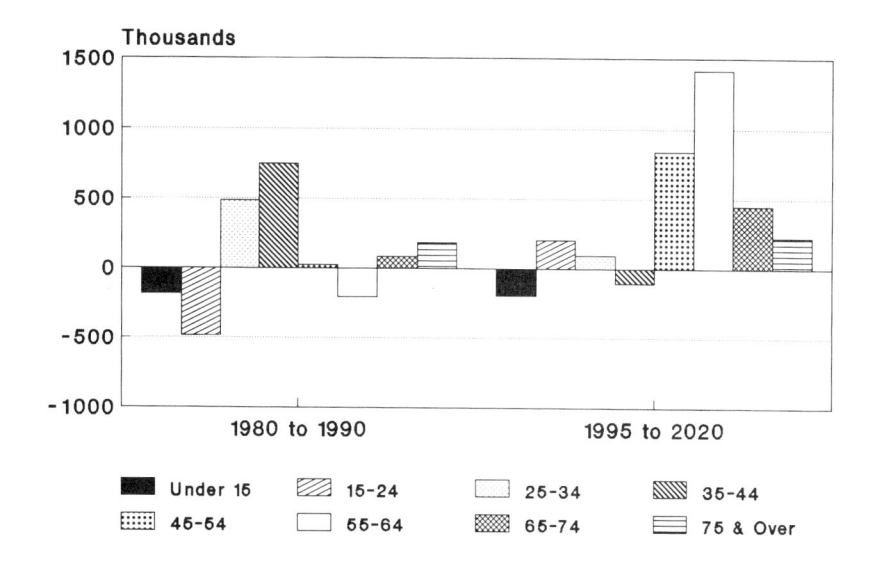

FIGURE 5. Changes in the Region's population by age 1980 to 1990 and 1995 to 2020.

force and demands of employers will persist over the long term, particularly since global pressures are exacerbating these conditions.

New York City

The boroughs of New York will continue to gain population as the City attracts nearly two-thirds of the Region's foreign immigration. Dominicans, Chinese, and Jamaicans constitute the largest flows of a highly diverse stream that has rejuvenated older neighborhoods as ethnic communities take root and flourish in the outer boroughs. All boroughs are expected to gain population as the City grows by roughly 6 percent over the next 25 years.

Long Island

In contrast to the recent past, which witnessed virtually no year-round population growth on Long Island, more new residents will be added over the next quarter century than employment. Average household size is likely to increase in fully developed portions of Nassau and Western Suffolk as "empty nesters" retire and resettle. Investments in LIRR service will improve accessibility for the commuter. Although special ground water protection zones will restrict density in developing areas of Suffolk, a full build-out is expected to occur elsewhere and some second homes will convert to year-round living with the rise in telecommuting. Between 1995 and 2000, the population of Long Island is projected to grow 13 percent.

Mid-Hudson

The Hudson Valley has grown in population in the 1990s, despite its employment losses, partly because of the affordable housing and developable land it offers. Mid-range counties such as Orange and Putnam have experienced considerable development pressure from northward relocation which will spread to Dutchess and Ulster. Both Asian and hispanic immigrants are attracted to Mid-Hudson communities, including older cities in Westchester County, and telecommuting is taking hold. Between 1995 and 2020, the Mid-Hudson is projected to grow by 20 percent.

New Jersey

The fourteen counties of Northern New Jersey will attract half of the Region's population growth between 1995 and 2020, expanding by 20 percent or slightly ahead of the pace of job growth. With its recent and foreseeable development, New Jersey has become more self-contained, though commutation and other linkages to New York will continue to enfold New Jersey in the Region. New Jersey's development pressures will focus on the coastal counties of Middlesex, Monmouth and Ocean, and the Hudson River waterfront.

Connecticut

Population has not grown in the Connecticut subregion in recent years and the future pace of population growth will likely be slower than other suburban areas in the Region. Immigrants have been attracted to Connecticut's older cities, but white out-migration has been significant and will likely continue as the economy changes. Development pressure is northward, away from coastal communities, into river valleys. Between 1995 and 2020, Connecticut is projected to grow by 11 percent.

ALTERNATIVE SCENARIOS OF DEVELOPMENT

While the baseline forecast driven by the perceived market-trends should be considered the situation that will likely obtain if no severe effects occur from global warming, it is useful to speculate about the demographic and economic factors that could contribute to alternative scenarios, as well as the sensitivities of socioeconomic conditions to greenhouse impacts. For simplicity, these ruminations are organized into: a *Sustainable Economic Future* with desired or optimal conditions, and a *Critical Uncertainties Future* with possible greenhouse effects.

Sustainable Economic Future

If the Region were to embark on a path toward sustainable economic development, several significant changes would result. For one, the value of natural resources, environmental quality and amenities—the so-called externalities of production—would be consciously incorporated into the price structure. Full cost accounting would mean that some forms of production would no longer be viable because of heavy environmental costs; other products and processes would be encouraged (such as automobile or appliance disassembly plants); material source, throughput, energy and waste reduction methods would dominate production processes; government operations and capital investment would be funded by "green fees" (such as congestion pricing); changes in consumer behavior would result in less material consumption; and leisure time would likely be valued more highly. Under this scenario, it is likely that more convergence would occur in the Region's income distribution as poverty is addressed directly by correcting price structures and new forms of work are developed.

Critical Uncertainties Future

Global climate changes with severe regional impacts in developing countries would encourage heavier immigration flows to the Region, placing greater stress on social support systems and labor force absorption. Greenhouse effects in the Region would divert capital needed for public infrastructure maintenance and private sector investment, contributing to greater inefficiencies in doing business and less cost-competitiveness in global markets. Significant losses would occur in the stock of natural resources and quality of life assets, discouraging in-migration of an educated labor force. Though most industries would not be particularly sensitive to global warming effects, other than by cost pressures, some segments of the population and settlement pattern would be directly affected. Older populations would be sensitive to heat stress, poorer populations to increased costs of living, and coastal housing sites to flooding, especially in Long Island and New Jersey. The probable outcome of critical climate and other uncertainties would be less income equality, lower per capita growth in income, and less economic well-being overall.

DISCUSSION OF THE PAPER

MALCOLM BOWMAN: It sounds like a pessimistic outlook. Is that how you would characterize this?

REGINA ARMSTRONG: Well, it definitely appears that the future will be much slower growing than was the past. There will be a widening of income disparity in the region. Without some sensitivity to the increasing environmental impacts that are occurring here, there will be a much lower quality of life. Certainly in some sectors of the economy, there will be much more productivity, but a split between the haves and the have-nots will occur.

FRED LIPFERT: I was struck by the frequent references to growth. I think I understand what it means, but in the context of this conference, is it fair to equate growth with more CO_2? Our economic future depends upon our own growth and worldwide growth. I would guess that means more CO_2.

ARMSTRONG: A lot depends not just upon the industrial composition of the region, which really is very much a service-oriented and information-oriented post-industrial economy, but upon factors that relate to it: how people spend their money, and I think it is in those sectors that the CO_2 is a factor more than in the direct production sector.

LIPFERT: My question really related to the global context. You indicated that we are strongly linked to global markets.

ARMSTRONG: Definitely, outside this region. I agree completely that our actions abroad and the economies that are being directed from headquarters here really are triggering that kind of an effect. And the products that people in other regions will buy may not be anywhere near as environmentally sensitive as what we attempt to sell close to home.

Waterfront Planning and Global Warming

HOLLY B. HAFF

*Transportation Division
New York City Department of City Planning
2 Lafayette Street, Room 1200
New York, New York 10007-1216*

*Robert F. Wagner Graduate School of Public Service
New York University
4 Washington Square North
New York, New York 10003*

INTRODUCTION

Waterfronts today are valuable untapped resources with great potential for reshaping our region. The vast stretches of shoreline that lie unused in the wake of industrial decline and transportation changes (ref. 1, p. 1) have created opportunity and challenge for the future.

Coastal and waterfront development became popular in the 1960s (ref. 2, p. 10) and has increased so that nearly 50 percent of the nation's population lives in the coastal area (ref. 3, p. 1). If global warming occurs, the ever-accelerating development at the shore will pose great risks from both the estimated increase in storms and hurricanes (ref. 4, pp. 1, 13, cover) and the rise in sea level (ref. 5, p. 12) with the related flooding and erosion. We are at a critical juncture in planning, caught between an increasing market for waterfront development, vast quantities of underused waterfront property, and Mother Nature's lessons of flooding, sea level rise, and global warming.

The questions for waterfront planners are twofold: What is likely to happen on the seemingly unlimited decayed waterfront stretches given the economic and political factors and, secondly, what are some of the ideas surfacing today that address sea level rise and global warming?

HOW THE WATERFRONT OPPORTUNITY AROSE

The legacy of these now unused waterfront acres started with the early European settlers who immediately recognized New York harbor as one of the best protected deep water harbors in the world and began to transform the natural shores of the region with development. Over time, pier structures and landfill were used to accommodate transportation and industry. Long finger piers were extended from bulkheaded shores where companies manufactured and warehoused goods to be shipped. By the mid-1800s, the region became the largest port in the country (ref. 6, pp. 3–9). The waterfront was highly developed, active with ships, tugs, wagons, and railcars. Port success continued through the 1950s with cargo movement and thousands of passengers using ocean liners.

In the early 1960s, shipping transportation technology began to make great changes in how the waterfront would be used. General cargo began to be transported via intermodal containers, which was easily loaded in bulk rather than breakbulk style—item by item. Centralized container ports were developed and, at the same time, passengers began using airplanes rather than ocean liners owing to the speed and cost advantages of air transportation (ref. 6, pp. 3–9).

These changes have had a dramatic effect on the waterfront. By 1970, many piers, wharves and docks were abandoned as goods began to be transported by trucks to the containerports. Waterfront property began to fall into disrepair. What was once an active shore became quiet and neglected. Despite the slow acceptance that piers were no longer needed, the underutilized waterfront properties began to reveal themselves as an opportunity to open up the harborfront for non-port uses. Increasingly developers, builders, and government agencies began to look to the waterfront—where properties were beautifully sited, had few relocation problems,[7] and were potentially enormous revenue generators.

Looking back over the past 25 years, it is difficult to imagine why, with so many development schemes and the abundance of available waterfront land, so little has been built to take advantage of the exciting waterfront opportunity. Notable successes include Battery Park City, Liberty State Park, South Street Seaport, development at Exchange Place, but scores of projects never made it. Why did so few plans that were on the drawing boards come to fruition?

HOW THE ENVIRONMENT AFFECTED WATERFRONT PLANNING

The answer is that waterfront redevelopment plans were on a collision course with both the new realm of environmental regulation and the complex array of socio-economic/political factors. Around 1970, a Federal environmental legislative agenda took hold. Nationwide legislation, such as the Water Pollution Control Act, the National Environmental Policy Act, and the Coastal Zone Management Act, was enacted providing the groundwork to address environmental issues. A new comprehensive waterfront planning approach incorporated the environment—and flood hazard areas—and began to accommodate the dynamic changes of both natural features and market trends. Rather than applying specific land uses to a site in the 1960s Master Plan approach, planners began to set out guidelines for appropriate growth with sensitivity for how much the land was capable of supporting.[8]

By 1980, planning efforts in New Jersey, New York and Connecticut were emphasizing the balance of economic development with public access and the protection of natural resources, in line with the Federal Coastal Zone Management Program. The 1980s development boom, therefore, became subject to the new environmental elements in discretionary reviews. Regulators and developers, who were unfamiliar with new, and sometimes more costly, regulatory requirements, struggled to mold formerly planned projects to fit the new environmentally oriented rules.

The true turning point in waterfront planning in the region was the defeat of the Westway project—a proposed underground highway built in new landfill. The project was killed in court over Hudson River fish habitat loss as analyzed in the environmental impact statement. This defeat brought waterfront development to an abrupt halt and set the tone for taking both the environment and community resistance seriously. Most developers became unwilling to risk the delays, opposition and lawsuits to in-water construction. Regulators, after losing in court, became more conservative—even timid—discouraging developers from in-water construction and shying away from waterfront projects as a whole. The scientific community struggled to identify clear standards to determine exactly when a development would adversely impact the environment. Community lawsuits on almost any environmental issue became a feared element in planning, and, as a result, greater natural feature awareness began to ease into the regulatory realm and the developers' consciousness.

ECONOMIC DEVELOPMENT AND POLITICAL FACTORS

Although the scientific community generally agrees that the sea level is rising and that there will be increased level of storms in this region either from natural cycles or from global warming,[9] valid issues have been raised in regards to the economic impacts of implementing the measures that some say are necessary to reduce global warming emissions (ref. 10, p. 1) or sea level rise. The general public, uneasy about the long-range nature of the predictions, raises the following questions for us:

- Who believes that global warming is something New Yorkers have to worry about?
- Why does the worry about global warming have to get in the way of "my" beach house?
- What's the best use of the waterfront and who gets to use it?

Believability

One of the greatest hurdles facing professionals who are responsible for waterfront development is the believability of sea level rise and resulting flooding to the general public. Based on the absence of any recent devastating storms, the public thinks it highly unlikely that a storm could damage their property. They simply do not believe that a hurricane of any significant force will hit this region.[11]

In addition, people generally do not understand the meaning of sea level rise. The atmospheric concept of global warming coupled with the longer-than-lifetime prediction is easy to dismiss as too complicated and irrelevant. When the estimated time that it would take the sea to rise one inch is beyond a few years, the impacts are not immediately felt and, therefore, are not significant enough to stir most individuals.

Even when people understand the concepts and the potential long range problems of the loss of lives and property, there is the tendency in the belief of technical solutions emerging in time to save the day. Professionals have a broader awareness of the costs of developing and implementing technical solutions and also know that some of these solutions have led to increased exposure or risk. The public relies on a belief that our high-tech society will come up with answers so that no one really has to worry about a storm, flooding, or erosion.[5]

Desire

Another hurdle to changing how development occurs at the shores is the strong attraction that water has for people. Even flooding and erosion risks have done little to deter people's desire to live and recreate near the waterfront and natural coastal areas. The 1990 coastal population in New York, New Jersey and Connecticut totals 16,791,074 persons—58 percent of the tristate total population—and the insured coastal property at risk is valued at a total of $533,438,323 in 1988 dollars.[12] The lure of the water appears to be magical—a powerful magnetic force that universally evokes fundamental emotions (ref. 2, p. 10). This desire is often so deeply rooted and communally shared that it can result in changing the minds of publicly elected officials even in the face of logical, well-researched plans.

In this regard, from a regulatory perspective, the Federal Emergency Management Agency's (FEMA) flood hazard regulations represent the best of both worlds: allowable development, but preventive controls. This is the subsidized program that provides flood insurance to property owners in flood hazard areas, providing that the owner builds to rigorous construction standards that are designed to withstand the wind and hydrodynamic pressures of storm and flood conditions and to protect human lives. In this way, developing in a hazardous area becomes the responsibility of the individual property owner.

On the other hand, allowing any development in flood-prone areas has had environmentalists and critics calling for the beach to return to a natural state.[13] While some government officials are careful not to encourage polar positions of environmentalists versus property owners, other state officials have made it clear that no compensation is due for lost homes on the coast.[14]

Theories that government should let nature take it course along the coastlines and encourage retreating from hazardous areas have been met with varying degrees of support. The National Park Service (NPS) is well known for its *laissez faire* approach to Mother Nature's wrath. NPS allows forest fires to burn in Yellowstone National Park and has watched well-used beaches erode and undermine park structures. Plumb Beach in Brooklyn was saved, not by NPS, but by New York State Department of Transportation to avoid losing the adjacent Belt Parkway to flooding and erosion.

Retreating from the shore will not come easily, if at all: witness Dune Road in Westhampton Beach, Long Island, where beach erosion destroyed 150 of 265 homes between 1992 and 1994. These losses came after years of erosion caused in part by publicly designed and built updrift groins—part of a broader erosion control project that was never completed. After 20 years of litigation and recent reluctance on the part of New York State officials, a settlement was reached in April 1994 in which the Federal, State, and County governments will rebuild—with taxpayers' money—the beach for $32 million, allowing property owners to rebuild their homes. The project also includes 30 years of beach maintenance.[15]

The emotional attraction and attachment to the coastal areas puts heavy pressure on elected officials. No official wants to tell people that they can not use the coastline because of potential hazards. A response from a town official when asked to "let nature take its course" typically can be predicted as the "tell that to Holland" routine. Images of the Dutch boy with his finger in the dike—holding back the sea—and the entire population cheering his bravery come out of our childhood culture, the reading of Hans Brinker.[16] Tax-paying property owners, who feel that their homes will be taken by government decisions, are angry and difficult for politicians to handle.

Highest and Best Use

Economically and politically, waterfront reuse translates into "who benefits?" and creates its own set of hurdles. "Who benefits?"—or who gets access to the property, for what uses and why?—are questions that often are sorted out through the political process.

The economic value of either the property or the use plays a major role in the ultimate decisions of "who benefits." Economics and politics become closely linked when assessing the uses and users and what they represent—either in income to the local tax base or in a constituency of voters.

Several industries vie for land along the waterfront. For example, leisure time recreation and tourism including swimming, sunning, boating, fishing, is an important expansion industry. Nationwide, tourism generated a payroll of nearly $80 billion and over $41 billion in tax revenues for Federal, state and local governments (ref. 2, p. 17). Tourism in Fire Island has attracted city dwellers for years, and the 1990 census data suggest that the mean property value is $450,000 in the village of Saltaire and $232,000 in the Village of Ocean Beach (ref. 5, p. 107). Waterfront redevelopment is often linked with tourism, as has been demonstrated at the Quincy Market in Boston or Baltimore's redeveloped Inner Harbor.

Opportunity to generate revenue by redevelopment of abandoned waterfront lands is staggering. Battery Park City's yearly income is over $90 million (ref. 17, p. 4), and will soon rise significantly with the 8,000-employee Commodities exchange about to be built. Numbers like these coming into government coffers ease the tax burdens of paying for all the basic services such as schools, police, firefighting, and garbage collection. From the bottom line of many successful waterfront projects, it would appear that residential, commercial or mixed use developments would be the highest and best use for the land.

However, regulators are often confused over how to redevelop the waterfront without foreclosing necessary industrial options and ruining the natural resources. There is a question as to how much water frontage should be given over to non-water-dependent uses. For example, the growing waste recycling and transfer industry may benefit from waterfront locations, such as moving tons of recycled newspapers by barge. Politicians are reluctant to move too quickly on land-use decisions that will have long-term effects.

In addition, although the waterfront land is often envisioned as having tremendous value, the decrepit nature of the property can create a marketing nightmare. Several years ago, the Port Authority issued a Request For Proposal (RFP) for acres of waterfront land along Brooklyn's waterfront with glorious views of the harbor, the Statue of Liberty and the Manhattan skyline. Although the RFP was distributed worldwide to attract redevelopment, the sites were sold at bargain basement prices to local family-owned businesses for manufacturing uses.

But, from a practical perspective, the political process lends itself to the redevelopment of underused and desired real estate. The realities of elections are that development and the highest and best use for vacant waterfront sites often impresses voters and the typically generous real estate contributors far more than technical discussions on how far the sea level may rise in the next 20 years. Four-year-term elected officials, under the gun to balance the budget for the next fiscal year, will find it difficult to give priority to planning for the prevention of global warming or the adaptation for sea level rise rather than a proposed waterfront development that brings in tax revenue.

In 1994, much of the formerly used industrial land along the waterfront remains unused and ripe for opportunity. Some blame the economic recession resulting from the 1987 stock market crash that followed close on the heels of the 1985 Westway decision as the reason for little or no development over the past 6 or 7 years. Others point to the hesitancy on the part of land-use decision-makers to rezone the waterfront, fear of environmental lawsuits, the battles over who benefits from reuse, and the insensitivity to the environment of the development community. Whatever the reasons, the remaining decrepit and underused waterfront is an opportunity for new approaches. These new approaches should allow for the inclusion of the prevention of global warming and adaptation should sea level rise result from an unavoidable greenhouse effect. In doing so, planners and developers will be protecting their properties for the future by avoiding flood inundation, loss of property and hazards to human life.

EMERGING RESOLVE

The scientific research, the growing environmental awareness, and recent storm flooding (FIG. 1) have not gone unnoticed by waterfront planners. New mechanisms are emerging to adapt waterfront planning to the potential dangers of global warming, possible increase in storms, and sea level rise. Owing to the difficulty that politicians have in building a constituency for addressing global warming, these mechanisms rarely are acknowledged as having roots in this very issue. They have often emerged through case-by-case planning, sometimes by individuals, sometimes by civic groups, sometimes by government pilot projects. These efforts, when taken cumulatively, begin to set policy that leads to more effective waterfront planning. They should not be underestimated, for they are precedent-setting and, with support, can become convention.

Private Beach Renourishment: The Fair Harbor Case

On the 32-mile barrier beach, Fire Island, damage was so severe from the December 11, 1992 Nor'easter and the March 13, 1993 blizzard that experts began to be concerned that the island could be breached with further storms. The Long Island southern shore potentially could be seriously affected—flooded—with significant property damage from Massapequa to Patchogue.[18] Fair Harbor home owners developed their own erosion control district, and through bonds they raised over $1 million to rebuild the beach and a 15-foot high by 50-foot wide dune.[19]

This self-imposed taxation of individual homeowners has alerted politicians not only to how important the issue is to each property owner. The debate has elevated to the broader issue of whose property—including public property—may be eroded and flooded out next.

Although this mechanism does not get at the heart of the problem, this "soft" form of erosion protection has far fewer impacts than structural methods. It places the cost more squarely on those who benefit from the beach location, thereby avoiding political hurdles.

Stiffening Regulatory Requirements Case by Case: Arverne

A second example of recent concern about flooding is government's role in the redevelopment of the low-lying peninsula known as the Rockaways. In the

1960s, a 300-acre urban renewal area, called Arverne, was bulldozed to be redeveloped. After many years of delays, 7000 units of housing were proposed in the late 1980s. The environmental impact statement (EIS) addressed coastal flooding and erosion by raising concerns about the low lying property (ref. 20, pp. II.J 17–21). As a result of the EIS analysis, the City and the State committed public funds to renourish the beach on an as-needed basis by paying half the costs. In addition, final approval required that the dune was to be enhanced and fortified, including landscaping to prevent overtopping.

But the most dramatic decision was to impose stricter development standards than established by the Federal Emergency Management Agency. The base flood elevation—that distance above sea level deemed to be safe for human occupation—was raised in the proposed development area so that all buildings would be subject to FEMA flood hazard A-zone construction standards.

In this Arverne case, government was wise enough to impose stricter standards in light of current theories on sea level rise, flood modeling, and knowledge of the local conditions, despite the increased building costs that these stricter standards would also impose.

Trails/Greenways/Esplanades: ISTEA's Unintentional Flood Control

Another mechanism that is indirectly addressing potential effects of global warming is the "greenway planning." Greenways are continuous linear accessways that create not only new park space, but alternative auto-free transportation routes for biking, roller-blading and walking. This Greenways agenda was given a boost in 1991 with the enactment of the Intermodal Surface Transportation Efficiency Act (ISTEA) that encourages cycling and walking as transportation modes (ref. 21, pp. 1, 6, 12). This boost was particularly important for urban areas where great opportunity on the waterfront existed and recession had limited its redevelopment. The waterfront greenways, such as those planned for the New York and New Jersey Hudson River waterfronts, provide an important setback area to reduce flooding impacts.

Renaturalizing Unneeded Bulkheaded Areas

The Regional Plan Association (RPA), which is currently writing a new master plan for the region, has suggested that environmental design should be one of the foundations of the revival of the industrial waterfront. RPA's ideal image is to reestablish the salt marsh estuary, whose wetlands have great capacity to absorb flood waters and may help to cope with the possible rising water level.[22] In New Jersey, a recent inventory by the Regional Plan Association found 2,500 acres of redevelopable coastal land in Union County, New Jersey. The county's plethora of waterfront property equals one-tenth of the land that would be needed for the estimated employment growth until 2010.[23] The RPA developed a model that encourages economic development and renaturalizing the shore.

Another approach that could be applied to renaturalizing the shore is the "urban forest" concept of planting trees in city streets to reduce energy consumption and absorb pollution to lessen the effects on global warming (ref. 24, pp. I–IV). The recent study in Chicago, measuring the environmental and financial benefits of trees, quantified how much the trees improve air quality and make it less costly

FIGURE 1. Flooding from the storm of 13 March 1993. **(a)** Exchange Place; **(b)** PATH subway entrance; and **(c)** Subway tunnel. (Photos courtesy The Port Authority of New York and New Jersey.)

FIGURE 1. (*Continued*)

to heat and cool buildings. The Chicago Urban Forest Ecosystem Project will be tested in New York City to see if street trees can reduce energy consumption.

Limiting Development: Waterfront Zoning

While the scientific community sorts out the specifics of sea level rise predictions, some level of concern about new development along the waterfront is appropriate. Again, not specifically in response to sea level rise but to comprehensive planning and environmental issues, waterfront zoning regulations were approved for New York City in 1993 which prohibit building piers or platforms for any use other than open recreation or water-dependent uses. In addition, the building on existing piers has been significantly limited to a height of 30 feet.[25]

CONCLUSION

These planning mechanisms provide guidance for future approaches to planning for potential sea level rise in this region where the waterfront is changing. Given industrial decline and the consolidation of shipping to containerports, it is clear that a significant portion of the decrepit and unused waterfront will become available for reuse during the next 25 years. For planners, it is a critical time to address how global warming prevention and potential sea level rise hazards can be incorporated into revitalization plans for the waterfront.

Generally, the waterfront planning goals for the region encourage the balancing of economic redevelopment with the preservation of natural resources, the accommodation of adequate working waterfront space, and increasing the public accessibility of the waterfront. The revitalized waterfront, in a region with hundreds of coastal miles, should be able to accommodate all these interests and address global warming prevention and potential sea level rise hazards. Below are some waterfront planning recommendations for consideration:

Global Warming Prevention

Plant the Waterfront to Improve Air Quality

There are 578 miles of waterfront in New York City alone. Many of the shores are devoid of any vegetation, but could be planted. For example, many active industrial properties that do not need ship access have desolate, eroding banks and bulkheads that could be planted and, at the same time would greatly improve the appearance of the industrial neighborhoods. Derelict industrial waterfront property often is difficult to market and investing in the renaturalizing of the shore could raise the value of the property and allow for greater visualization of the site for alternative uses. Areas of the region are also ringed with highways that could be planted with more trees. Vegetation is relatively cheap and has relatively low maintenance, as long as native and coastal species are used.

Renaturalize Unneeded Bulkheaded Shorelines

Miles of bulkheaded shoreline exist without reason now that shipping is consolidated in containerports. Renaturalizing the waterfront would help to absorb rising waters and could also improve air quality.

Sea Level Rise Adaptation

Design Greenways and Bikeways That Help to Prevent Upland Flooding

Greenways along the waterfront should be aggressively supported in that their built-in dual purpose is a setback or buffer, protecting the uplands from flood waters. With forethought, the design of these greenways, particularly with the help of transportation funds, could include upland flood control. Should sea level rise predictions warrant it, the greenways could be built with multiple levels and should be assessed for sea wall potential.

Arm Politicians with Ideas for the Next Storm

Although the voting public typically may not find discussions about sea level rise appealing, the exception to this situation is emergency response. Emergencies created by storms and resulting flooding and erosion catch media attention. TV coverage of swirling winds, crashing waves, collapsing piers, and neighborhood families being evacuated by authorities are images that move the public. At these times, voters want to know what government is doing to prevent such disasters in the future.

This is the time to recommend national standards to address sea level rise. Clearly, the level of the sea is not a local issue. Increased regulation could come in the form of FEMA adjusting its formula to raise the base flood elevation.

The Hurricane Evacuation Computer Model for the State of New York offers unique visuals to understand the flooding conditions. Changing the level of the sea in the model to assess potential flood hazards would allow a comprehensive understanding of the impacts during various storm conditions[26] and add to public support for increased flood protection.

Encourage Interim Uses in Transition Areas

Where it is unclear what the future of the land use of an area should be, loosen the regulations to allow for a greater variety of activities. In particular, publicly held lands that are being held for possible future maritime uses could be reused for low-cost commercial recreation. Idle now, many public piers and waterfront properties are deteriorating and adding to the maintenance costs for future uses. It would be beneficial to put these properties back in use where capital investment costs are not great and maintenance costs are absorbed by the users.

Incorporate Flood Hazards into Zoning

To date, most localities use their building codes to implement flood hazard regulations. However, it is possible that, if the predictions of sea level rise were severe enough, local planning commissions would undertake controls that would limit new development to areas on higher ground. If a well-considered plan is adopted, the legal challenges to restricting private development on private property may be upheld.

This approach costs less than building sea walls or other structures and avoids both the false sense of security and the possible additional hazards that sea walls impart. When walls are breached, they can trap retreating water thereby lengthening the time structures are under flood waters. And the walls, when breached, act to increase the velocity of incoming water.

What must be considered, however, is the loss of income from development potential of low-lying areas and the estimate of risk of a hazard occurring. Are the costs of disastrous floods greater than the income and pleasure derived from the development of coastal property, when other environmental impacts can be avoided or mitigated? Or are the risks to human life great enough that dollars should not be a significant factor in the decision?

Survival at Stake? Concepts for Building and Retreating

In the worst case scenario, choices narrow to retreating from the shore for residential or commercial uses or to building protective structures.

The concept of retreat is a difficult one for a dense metropolitan area. The level of public investment and settlement is so significant that abandoning these areas becomes impossible. The costs of buying out thousands of individuals, businesses, corporations, and institutions are beyond the state's or locality's budgets. Unless storm destruction was of such an enormous magnitude, it would also be difficult to imagine convincing residents to leave the area. Although some sparsely settled areas of the region may be lost, the concept of retreat is highly unlikely and focuses attention on structural solutions.

In recent years, the use of structures to prevent flooding and erosion has been discouraged for environmental reasons. However, with survival at stake, a system of flood gates, sea walls, and if necessary, locks is not inconceivable. The reasons that have made New York City a successful harbor—its highly protective nature—are the same reasons to believe that an engineering solution can be accomplished by blocking the sea at the three narrow entry locations: the Narrows, the Arthur Kill, and at either the Whitestone or Throgs Neck area. This would protect the denser, more developed areas of New York and New Jersey. The critical factor for walling off the harbor would be accommodating the active shipping industry through gates or locks. Other locations, such as Coney Island and the Rockaways, would require some type of sea walls. These areas may warrant protection due to the high level of public investment in infrastructure and the relatively high density. Other areas, such as Jamaica Bay, may not need as much protection as imagined owing to buffer areas and the airport which has an enormous amount of land, but most of it vacant or improved with only runways or airplane taxi areas.

―#―

The New York/New Jersey metropolitan area has limited ability to solve the global warming problem. Waterfront planning has even greater limits in what it can do. However, it is important to ensure that the enormous opportunity that lies at the water's edge for redevelopment be taken within the framework of both preventing global warming and addressing the potential of sea level rise. The balance must be carefully weighed to address environmental, socioeconomic and political factors. The greatest challenge is for this global city and its region to proactively initiate and implement techniques to allow for new environmentally sensitive development and multiple uses on our waterfronts in full awareness of global warming.

REFERENCES

1. DEPARTMENT OF CITY PLANNING. 1992. New York City Comprehensive Waterfront Plan: Reclaiming the City's Edge. August.
2. BREEN, A. & D. RIGBY. 1994. Waterfronts: Cities Reclaim Their Edge. New York: McGraw-Hill.
3. CULLITON, T. 1990. Fifty Years of Population Change along the Nation's Coast: 1960–2010. U.S. Department of Commerce. April.
4. TAIT, L. S. 1992. Coastline at risk: The hurricane threat to the Gulf and Atlantic states. Excerpts from The 14th Annual National Hurricane Conference, April 1992. Federal Emergency Management Agency, Washington, D.C.
5. COASTAL EROSION TASK FORCE. 1993. Draft final report, Volume Two: Long Term Strategy. Department of State and Department of Environmental Conservation, State of New York. November.

6. Moss, Mitchell. 1980. Staging the Renaissance on the Waterfront. New York Affairs, New York Univeristy.
7. Department of City Planning. 1970. Waterfront Supplement to the Plan for the City of New York. City of New York.
8. Department of City Planning. 1974. Comprehensive Waterfront Planning Report. City of New York.
9. Hill, Douglas. 1994. The Baked Apple? Metropolitan New York in the Greenhouse. Infrastructure Group, Metropolitan Section, American Society of Civil Engineers.
10. Ginsberg, William R. 1994. The Threat of Global Climate Change—What Can New Yorkers Do?: State and Local Strategies to Reduce Greenhouse Gas Emissions in New York State. Report of the Environmental Law Section of the New York Bar Association. January.
11. New York City Mayor's Emergency Control Board. 1994. Hurricane Contingency Plan. BM 181-G (Rev 4-94): 1–27, New York.
12. Tait, L. S. 1992. Coastline at risk: The Hurricane Threat to the Gulf and Atlantic States. Excerpts from The 14th Annual National Hurricane Conference, April 1992. Federal Emergency Management Agency, Washington, D.C.
13. Shellback, Peter. 1994. Newest Village Is a Ghost Town. The New York Times, January 23, Sec. 13LI, p. 6.
14. Nordheimer, Jon. 1993. Exposing Oceanfront Property and Human Folly. The New York Times, January 24, Sec. 1, p. 25.
15. The New York Times. 1994. Pact Reached in Beach Erosion Project. April 24, p. B6.
16. Green, Norma. 1974. The Hole in the Dike: Hans Brinker and the Silver Skates Retold. Scholastic, Inc.
17. Ernst & Young. 1994. Financial Statements: Battery Park City Authority. Battery Park City Authority.
18. Falk, William B. 1993. Doomsday for the dunes? On Fire Island, it's all wasting away. Newsday, May 17, p. 5.
19. Johnson, Bill. 1994. Fair Harbor homeowners' experience in beach erosion and developing and erosion control district. Speech to the New York University Summer Institute, Planning and Management of Urban Waterfronts, June 29.
20. Department of City Planning. 1990. Arverne Urban Renewal Project: Final Environmental Impact Statement. City of New York, June.
21. Department of City Planning. 1993. A Greenway Plan for New York City. City of New York, Fall.
22. Yaro, Robert D. 1992. Working Paper Number 15: Visual Simulations of Future Development of the Tri-state Region. Regional Plan Association. May, 15B–15C.
23. Morgan, L., L. Hustourn & P. Henry. 1994. Union County Model Site Redevelopment Project Final Report. Regional Plan Association, June.
24. McPherson, E. G., D. Nowk & R. Rowntree. 1994. Chicago's Urban Forest Ecosystem. Radnor, PA: U.S. Department of Agriculture.
25. City Planning Commission. 1993. Special Regulations Applying in the Waterfront Area. Zoning Resolution. Department of City Planning, City of New York. Article VI, Chapter 2, pp. 361–414.
26. Federal Emergency Management Agency. 1993. A Hurricane Evacuation Computer Model for the State of New York: Documentation and User's Guide, April.

DISCUSSION OF THE PAPER

Lawrence Kleinman: The pictures that you showed certainly indicated that we have a flooding problem now. If we have sea level rise, it will get somewhat worse. Can you put that in perspective, how much worse it will be because of global warming?

HOLLY HAFF: The best way to take a look at that is to get ahold of FEMA's software: the hurricane evacuation computer model for the State of New York. It has great images, good maps. It shows you what areas would be inundated in 1, 2, 3, 4 and 5 level storms. You really get a sense of what would happen. The one thing that software lacks is the ability to change the level of the sea. It would be exciting to see them incorporate that into their program.

Global Warming, Infrastructure, and Land Use in the Metropolitan New York Area

Prevention and Response

RAE ZIMMERMAN

Robert F. Wagner Graduate School of Public Service
New York University
4 Washington Square North
New York, New York 10003-6671

INTRODUCTION

The metropolitan area of New York, New Jersey and Connecticut is surrounded by an extensive shoreline, portions of which have the highest population density in the country. The shoreline at the same time affords amenities and vulnerabilities to natural and man-made disasters, such as the potential perils of global warming. Predicted changes in sea level attributable to global warming may claim much of the waterfront areas that have been redeveloped for public use over many decades. Changes in shoreline populations and economic activity induced by global warming necessarily affect the entire metropolitan area.

Infrastructure plays a critical role both in protecting existing human settlements and in changing settlement patterns to reduce their contribution to global warming in the future. This paper addresses the vulnerability of infrastructure to the projected or estimated conditions (primarily sea level change) associated with global warming and the extent to which infrastructure can be modified to reduce its vulnerability and consequently, that of the land uses it serves (see Miller[1]).

Two approaches or scenarios are addressed. 1) In the short term, measures can be taken that are more responsive and reactive. That is, given that global warming will occur, those measures would reduce the likelihood that the immediate consequences of global warming would be felt by the area's population and economic activity. 2) Over a longer time period, the key issue is whether infrastructure changes such as relocation, redesign, retrofit, or renovation can be accomplished in a way that reduces the contribution of infrastructure and land use activities to global warming. Examples of such reduction strategies are energy conservation and/or emission reduction. In addressing these two perspectives, the focus will be upon infrastructure and its relationship to land use. Changes that might occur in the area's land uses directly in reaction to global warming (not associated with the services the area's infrastructure provides) or for other reasons, are not addressed here.

These approaches or scenarios operate in the context of anticipated impacts or consequences of global warming. Therefore, before addressing the infrastructure and land use dimensions, the impacts of global warming (in terms of sea level rise) that are assumed will be briefly set forth.

POTENTIAL CONSEQUENCES OF GLOBAL WARMING FOR INFRASTRUCTURE AND LAND USE

The effects of global warming on the NYC metropolitan area's infrastructure are largely a function of several interrelated conditions projected to be associated with the warming phenomenon: flooding, high winds, and high temperatures. The focus here will be on one of the major effects: increased flooding and submergence of low-lying areas from sea level rise. Sea level rise is in turn ascribed to a variety of climatic factors related to higher temperatures. These factors include higher levels of precipitation (from increased cloud formation), storm surges from more frequent and severe winds of hurricane force, thermal expansion of ocean waters due to increased temperature, and melting of ice caps. Some of these phenomena have already been observed. Others have been estimated.

Global estimates of both sea level rise and temperature changes are given in TABLE 1. The table shows a wide range of estimates that exist for sea level rise with best estimates of 18 cm (.6 feet) by 2030 and 44 cm (1.4 feet) by 2070. the Intergovernmental Panel on Climate Change (IPCC) report gives slightly higher best estimates of a 20 cm (.7 foot) rise in sea level worldwide by the year 2030 and a 1 m (3.3 foot) rise by 2100 (Bird,[2] p. 2; IPCC[3]).

Sea level rise is not a new phenomenon in the metropolitan area. Studies conducted for the Marine EcoSystems Analysis (MESA) project for the New York Bight project pointed out that sea level along the East Coast of the U.S. has been rising at least since 1895. Moreover, according to MESA, sea level rises over the past 20,000 years have been considered the cause of the submergence of the mouths of both the Hudson and Raritan Rivers (Yasso and Hartman,[4] pp. 10, 34). If this is correct, then global warming could result in either a continued upward trend in sea level rise or a sharper rate of increase.

Although regional levels are difficult to quantify, qualitative modifications in global estimates of sea level change shown in TABLE 1 can be made for the metropolitan area based on the area's environmental characteristics that might be expected to modify sea level rise. Two factors that can be expected to modify the global estimates are tidal conditions and erosion and sedimentation processes.

First, sea level rise in the metropolitan area could influence the tidal conditions. According to Bird[2] (p. 27), in embayments with narrow openings, sea level rises are likely to make tidal ranges higher in the interior portions of the Bays. In more open waters, changes in tidal ranges are likely to be lower. Moreover, Bird points out that, where the length of the tidal wave

TABLE 1. Initial Conditions Associated with Global Warming

	2030		2070	
	Range	Best Estimate	Range	Best Estimate
Temperature				
Global	0.7 C–1.5 C	1.1 C	1.6 C–3.5 C	2.4 C
Local-winter	1.5 C–3.0 C		3.0 C–6.0 C	
Local-summer	1.0 C–2.0 C		2.0 C–4.0 C	
Global Sea Level	8–29 cm	18 cm	21–71 cm	44 cm

NOTE: Mintzer[30] (p. 446), citing the World Meteorologic Organization 1985 report, gives an alternative figure of a 40–120 centimeter rise in sea level worldwide as a result of a 1.5–4.5 C change in temperature.
SOURCE: IPCC[3] and Broccoli.[29]

equals that of the Basin, resonance occurs and increases in the range are likely to be higher still. Thus, rises in tidal ranges in the New York Harbor Area are likely to be higher than global averages, given the narrowness of the points of entry of water into the Harbor (the Narrows and the East River) relative to the area of the interior Harbor waters. In addition, near resonance conditions that exist in Long Island Sound are likely to produce still greater variability in the tidal range in the harbor.

Second, patterns of erosion and sedimentation can either enhance or offset the effect of sea level rise on shorelines. According to Bird[2] (p. 61), the major general effect of sea level rise on coastal estuaries is considered to be a widening and deepening of the estuary, as well as an increase in salinity. The metropolitan area coastline is a candidate for such impacts, though the extent of such changes is difficult to predict with currently available models (Bird,[2] pp. 56–57).

In sum, a major estimated impact of global warming is a rise in sea level of 2½ to 3 feet, which is expected during the latter part of the 21st century. Taking into account local conditions, this estimate is likely to be somewhat higher in the metropolitan area.

Thus, for the purposes of evaluating impacts in the metropolitan area and for planning purposes an estimate of about **4 feet of sea level rise** may not be unreasonable given the existing projections and local factors that might alter global estimates in the metropolitan area. These elevations are necessarily highly uncertain, and are assumed for planning purposes only.

In applying the 4 foot estimate as a criterion for vulnerability of infrastructure and land use, it should be kept in mind that such a level raises all flood elevations by 4 feet. In other words, flooding episodes not related to global warming could have more severe consequences because of the overall higher water level.

A few other factors are important as well in determining the geographic dimensions of global warming, such as the location of flood waters within the metropolitan area and the factors that contribute to the severity and direction of the impacts.

Once a level of sea level rise is assumed, the elevation of the coastline or shoreline is needed to determine where flood waters would hit inland. This is not an easy task, since, especially in the metropolitan area, methods of arriving at shoreline elevation estimates vary, though some conventions exist. One convention is the uniform adoption of a 5–10 m contour line above sea level or a 1 km distance inland (Bird,[2] p. 5). Many metropolitan area jurisdictions use a larger distance for the purposes of coastal zone management. Delineation of flood hazard areas on flood plain maps can provide still another, but more direct basis for locating the extent of flooding. Their accuracy, however, depends on whether or not they have been updated sufficiently.

The effects of flooding due to sea level rise on infrastructure and land use vary according to the speed and height of floodwaters and their duration as well as the degree, rate, and pattern of sediment movement associated with the flooding. The rate, in particular, that is as significant as the magnitude of the rise. Currently, estimates of sea level rise globally are 1–2 mm per year (Bird,[2] p. 3).

Many of the effects of flooding on infrastructure and land use induced by global warming that are discussed below can be extrapolated from existing flooding episodes in the metropolitan area and flooding episodes elsewhere that have specifically affected infrastructure facilities.

LAND USE AND INFRASTRUCTURE RELATIONSHIPS

Interdependency of Land Use and Infrastructure

Any changes in the design and location of infrastructure made in response to sea level rise are likely to have extensive impacts on land use. The ability to quantify and even to conceptualize these interrelationships and hence the impact of global warming on land use and infrastructure is limited by the paucity of reliable models and the high degree of uncertainty in the nature of initiating conditions and system behavior[a] Ortolano,[5] for example, identifies two extreme methods to measure land use and infrastructure interactions: the large scale model and expert judgment using the Delphi technique. He concludes that neither one is sufficient to characterize land use/infrastructure relationships. Moreover, the population and economic forecasting component of large-scale models is a major source of uncertainty that increases model error. Even though the value of current models in predicting large scale land use changes from sea level rise is limited by the uncertain relationships between flooding and land values, with infrastructure acting as an intermediary in this relationship, some of the general relationships and concepts may be of use. A number of small scale models exist, for example, that allow an identification and quantification of some of the more obvious interactions under both response and prevention scenarios. The most popular of these are property value models. These models portray the attractiveness of land use in terms of the value of land. Land value is then predicted in terms of a variety of factors, among them, the availability of infrastructure services. It is well known that property values are often affected by the location of infrastructure. Proximate locations to infrastructure such as power plants, wastewater treatment plants, and roadways tends to depress property values. The provision of the service, however, tends to increase property values.

Some scenarios that reflect the broad concepts embodied in these models are proposed, even though detailed modeling of this behavior is beyond the scope of this paper.

Scenarios for Infrastructure and Land Use Interrelationships under Conditions of Sea Level Rise

Scenario 1. Response

A scenario that is reactive or responsive to sea level rise involves full prevention of the consequences of sea level rise largely through in-place infrastructure redesign, construction, maintenance and operational controls. The expectation here is that land use is to be protected by infrastructure. Under this scenario, land use patterns would remain virtually unchanged; that is, they would continue in the

---

[a] A profusion of large-scale models were developed in the late 1960s and 1970s in response to the environmental, energy conservation and associated growth management movements, and the need to address impacts of population growth. The large uncertainties associated with the models and the data inputs led to a decoupling of them, and a shift to smaller and simpler models. These had the advantage of greater control over discrete parameters and ease of comprehension at the expense, however, of less of an understanding of intervening and confounding factors that had to be eliminated.

direction they would have taken if sea level changes had not occurred, since acceptable protection is assumed.

Scenario 2. Leveraging Infrastructure and Land Use to Respond to and Retard Global Warming

A scenario that aims at eliminating sources of global warming within the NYC metropolitan area as well as preventing its impacts, could involve withdrawal of waterfront uses to higher ground for protective reasons plus greater concentration of activities in the metropolitan area for the purpose of reducing emissions from stationary and mobile sources of air pollution. Adaptation of infrastructure to sea level rise could either be minimal, be similar to that which was described under *Scenario 1*, or be substantial, involving land uses that are abandoned and relocated in upland areas rather than being altered in-place to adapt to sea level rise. Density would be expected to increase since available land for development would decline as a result of submersion (assuming that there isn't massive relocation out of the Region which would open up more land for redevelopment).

SCENARIO 1: RESPONSE

Infrastructure in the Metropolitan Area and its Water Dependency

In the metropolitan area, New York City's shoreline alone is 578 miles (NYC Department of City Planning,[6] p. i). This represents close to one percent of the total U.S. shoreline. A large portion of this shoreline is occupied by water-dependent infrastructure. Quantifying water-dependent infrastructure is precluded by the paucity of infrastructure inventories in general, much less those which identify infrastructure vulnerable to environmental hazards by virtue of its water dependency. There is no single measure of all of the existing infrastructure stock, though estimates for portions of it are available from individual agency data.

What is clear is that a considerable amount of infrastructure either occupies waterfront location or is vulnerable to flooding owing to its low elevation. Some examples within the New York City alone are as follows (Wagner,[7] p. 27 and updates as indicated):

- Forty-seven waterway bridges span the waterways within and adjacent to NYC.
- The wastewater treatment and collection system encompasses 6,100 miles of sewers which feed into 14 water pollution control plants with waterside locations, 80 sewage pumping stations, and 450 combined sewer overflow regulators that empty into NYC waterways.
- The distribution lines of the NYC water supply system consist of 32 million linear feet of trunk and distribution mains, controlled by 20,000 trunk valves.
- There are 6,375 miles of streets (NYC Department of Sanitation,[8] p. 2-1) and 2,057 highway bridges and elevated structures.
- The subway portion of the transit system encompasses 648 miles of mainline track available for passenger service which extend over 230 route miles. Of the 648, 411 track miles (or 137 route miles) are underground, 174 track miles (70 route miles) are elevated, and 63 track miles are at grade or are open cut (Metropolitan Transportation Authority,[9] p. 18). The system also includes 43 bridges and 213 power substations. The system operates 343 pumping

stations which remove an average of 15 million gallons of water a day accumulating from rainwater, high water tables, and water main breaks.

The New York City portion of the metropolitan area disposes of thousands of tons of municipal solid waste (MSW) per day, largely originating from within the City's borders.[b] Many solid waste management facilities have waterfront locations, given the unique geography of the metropolitan area with its many miles of coastline and the historic use of waterways as a means of transporting waste material. Landfilling, for example, is the major form of disposal of municipal solid waste, and landfills currently in use as well as those that are being closed often occupy waterfront locations or are located at sea level, extending over several thousand acres in NYC. The Fresh Kills landfill on Staten Island alone occupies 2,900 acres (NYC Department of Sanitation,[8] p. 3-25). These have traditionally been located near waterways given the reliance of solid waste transport upon barges. Marine transfer stations throughout the metropolitan area are major handling and distribution points for solid wastes. There are 8 Department of Sanitation stations, all of which are on the water, and 115 private stations, only some of which are on the water (NYC Department of Sanitation,[8] p. 3-6).

Much of the energy production infrastructure in the area is water dependent. Close to 80 electric power generating units in the Tri-State area (but for fewer actual plants), owned by about a dozen different companies, all have coastal or riverine locations, given their historic dependency on water for the delivery of fuel, as a source of water for cooling systems, and for the discharge of wastewater. (This area includes southern New York State extending south from Greene, Ulster, and Sullivan counties west of the Hudson and south of Dutchess county east of the Hudson.) Several of the generating facilities are nuclear, including Indian Point, Connecticut Yankee, and Oyster Creek, in total comprising six separate units.

There are numerous terminals for oil refineries and associated products. Tank farms are numerous. Dikes designed to prevent oil seepage to adjacent waterways when a tank leaks, presumably could withstand floodwaters from undermining the tanks.

The Location of Vulnerable Infrastructure

The ways in which flooding could generally impair existing infrastructure include:

- Malfunctions, since much of the infrastructure, even though it requires waterfront locations, cannot function under permanently submerged conditions.
- Material impairment due to erosion and waterlogging from continuous inundation and persistent submersion and corrosion due to the infiltration of subsurface infrastructure with salt water.
- Structural impairment due to the weakening of physical support structures.

Temperature increases, quite apart from flooding, create still another set of effects. Structurally, much infrastructure is not built to withstand extremes of temperature. Functionally, extreme temperatures could create changes in usage patterns, for which infrastructure is not designed.

---

[b] The NYC Solid Waste Management Plan Final Generic Environmental Impact Statement estimated that 11,350 tons of solid waste were considered municipal solid waste, and were collected by the City. Another 14,000 tons were collected by private carters from private businesses. (NYS DOS[8], p. 1-1).

Transportation

A comprehensive set of data tables showing the vulnerability of transportation infrastructure in the Metropolitan area by virtue of location above sea level was compiled in June 1994 by transportation agencies under the direction of the U.S. Army Corps of Engineers (USACE).[c] Some of the patterns of vulnerability of infrastructure to flooding suggested by that data are described below.

Bridges and Roadways[d]

a. Locational Vulnerability

According to the USACE survey, bridge access roads are the most vulnerable to the immediate effect of flooding. Although many of them are above the 4 foot increment, they would still be vulnerable to the storm surges and floodwaters that would now be higher, since they would be added to the rise in sea level.

b. Hydraulic Failure: Bridges

A major impact of flooding on smaller bridges located in the more rural parts of the Metropolitan area is washouts. Small bridges are usually not designed to resist hydraulic failures associated with high water velocities or increased water pressures due to high volume. Bridges that are more susceptible to such failures are typically older bridges, built according to standards that did not take into account such sources of structural failure. Although there is an extensive inventory of bridges and their condition throughout New York State (the Bridge Inventory and Inspection System) and New Jersey as well, which includes these local bridges, no systematic record of local bridge failures exists. Local governments usually rebuild these at their own cost at lower standards, and take financial responsibility for the risk of failure.

The probability of structural failure of waterway bridges during flooding conditions is a function of the velocity of water. Initial flood waters due to sea level rise are usually not the source of the highest velocities, since sea level rise would occur slowly. However, as the volume of water in a channel increases as sea level rises, its volume and speed would increase as a result of additional waters originating from the normal periods of spring thaw and flooding.

c. Corrosion of Roadbeds

The higher salt content of water inundating roadways and bridge decks due to sea level rise in an estuarine environment increases the corrosion of reinforcing steel. This could result in the removal of the top and bottom portions of the

---

[c] This database was collected to estimate the vulnerability of transportation infrastructure to a "Class 3" hurricane, with a velocity of up to 130 miles per hour. In addition to the wind impacts, the data base gives flood elevations expected from such a hurricane. Such a hurricane is estimated to have a maximum storm surge elevation rate rise of 13 feet per hour and a peak of 24 feet. Although these particular conditions are not necessarily applicable to global warming effects, the elevations for infrastructure contained in the data base are useful for identifying those structures that would be most vulnerable to a permanent rise in sea level. The complete report has been published by the U.S. Army Corps of Engineers.[10]

[d] This section is based on a telephone interview with Bill Winkler, NYS DOT, 9/12/94.

concrete on bridge decks. Standing water that is saline would increase the problem. Corrosion is not a problem for bridge footings, since the steel is located further inside the structure. A mitigation measure that is being used to make steel chloride proof is to put an epoxy coating around the steel. The effectiveness of such a coating depends on how it is applied. Any flaws that allow openings in the coating will reduce its effectiveness.

d. Heat

Bridges are designed to withstand a range of temperature change of about 120°F., from −20° to 100°F. To the extent that global warming conditions raise temperatures above 100°, road surfaces will be stressed. The tolerance of the surfaces would also be reduced, a phenomenon which happens even now. When highway expansion joints become filled with sand and debris, less movement is tolerated and the blockage of the joints prevents the surfaces from expanding as temperatures rise. The movement that results is called a "blow up." The problem increases with increasing temperature. It is easy to correct the situation through maintenance; however, persistent high temperature conditions would produce a continuous stress on the surfaces in spite of attempts to keep the expansion joints clear.

TABLE 2. Selected Vulnerable Points for Transit Facilities Based on Elevations, New York Metropolitan Area, 1994 (USACE[10])

A. NYC Transit Facilities: Station Entrances and River Tunnels

Feet below NGVD[a]	Station Entrances	River Tunnels
0–3.0		
3.1–6.0		
6.1–9.0	2	3
9.1–12.0	3	6
>12.0[b]	8	2

[a] NGVD: National Geodetic Vertical Datum of 1928.
[b] Those identified as being vulnerable to a Class 3 Hurricane.

B. Top of Rail Elevations at or Below Sea Level (feet below NGVD[a])

Penn Station	−6.3
PATH Stations	
Exchange Place	−71.0
Pavonia	−43.0
World Trade Center	−20.3
Grove Street	−15.6
9th Street	−15.0
Christopher Street	−14.6
Hoboken	−12.0
12th Street	0.0
Metro-North	
IRT Steinway Tube	−14.0

SOURCE: This information was derived from the data in the report: U.S. Army Corps of Engineers, FEMA, National Weather Service, and NY/NJ/CT State Emergency Management,[10] Metro New York Hurricane Transportation Study, Interim Technical Report, November 1995.

Transit Tunnels and Facilities

The points of the transit system that are most vulnerable to flooding are tunnel air and vent shafts, station entrances that are close to sea level, and berms that currently protect transit tracks, tunnels, and track. Some of the more sensitive points in the system identified in the USACE tables are tabulated in TABLE 2. The PATH system has many stations that are located far below sea level.

Some rail lines are actually located below sea level in a number of places, shown in TABLE 2B. The highest point in the MTA's subway system is the Smith and G Street station in Brooklyn, which is 88 feet above street level. The lowest point is the 191st Street station which is 180 feet below ground surface. The USACE compilation indicates that the lowest "top of rail" elevation above sea level occurs on the IRT 4, 5 & 6 lines where the elevations average 8 feet. Most of the lowest elevations are found in lower Manhattan.

Airports

Elevations of the major airports in the metropolitan area are shown in TABLE 3.

La Guardia is the one major airport that maintains a dike and pumps for floodwaters. The other two major airports have higher elevations and operate without such facilities. The airports in general are limited in the height of dikes that can be tolerated, because the angle of descent for aircraft can't be too steep as a result of the presence of dikes of higher elevation.[e]

TABLE 3. Elevations of Major Airports in the New York Metropolitan Area

	Elevation[a]
Teterboro	5.0
La Guardia	6.8
Newark International	10.3
John F. Kennedy International	11.7

[a] Feet above National Geodetic Vertical Datum (NGVD) of 1928.
SOURCE: U.S. Army Corps of Engineers.

Port Facilities[f]

The extent to which the functioning of port facilities will be impaired by rising sea levels depends on the type of docks that are used, and how flexible they are. For passenger service this refers to whether the docks are fixed or floating. Floating docks are more flexible, and their height adjusts to changing tides, although ultimately there is a limit to which they can rise and fall. For the docking facilities between Hoboken and the World Financial Center, the Port Authority operates floating docks. Some ferries use floating facilities as well. For example, the Staten Island facility uses a floating bridge to the boat, which is adjustable, even though

---

[e] Interview with Paul Wood, Port Authority, 1994.
[f] Interview with David Frainer, Port Authority, 1994.

the terminal facility is fixed. This is a common adaptation to tidal cycles. New York City has adapted to tidal variation for years, though the variations are not as extreme as might be projected for global warming.

The vulnerability of port freight facilities depends upon the type of technology used to transfer freight from vessel to dockside. Ports such as Elizabeth, New Jersey, operate with lift-on/lift-off facilities. This is the most common type of technology. The docks remain stationary, and container boxes are removed with cranes, stacked at dock side, and put on rail cars for further distribution. Higher sea levels do not present a problem for this type of technology because the container or cargo is not moved from the vessel to dockside directly; it is lifted. The system is not a problem as long as there is a sufficient margin built into the design of the cranes to adapt to changing sea levels. It is the roll-on roll-off facilities that are a problem, but these are rare. Moreover, even for these facilities, the ramps can be lengthened to adjust to increased sea level. A large margin is already built into the design of freight-hauling facilities, so that a rise in sea level would just reduce the margin somewhat.

Water Pollution Control Facilities

The major water pollution control facilities that would potentially be affected by a permanent rise in sea level are the water pollution control plants, combined sewer outfalls, and tide gate/regulator assemblies.

By virtue of their function, most water pollution control facilities in the New York Metropolitan area (as elsewhere) are located on or very near the water. Others discharge into and are hydraulically connected to waterfront facilities. Wastewater facilities in the Metropolitan area represent billions of dollars of public investment. Most of them are made up of rigid structures not easily elevated. Elevation of structures or extensive pumping and sea wall construction to prevent inundation would be required to secure the facilities against the effects of sea level rise. The combined sewer system in areas which are prone to tidal flooding is part of a system of tide gates and regulators. The practice of combining household waste with rainwater dates, according to Tarr, from 1845 when the City Council permitted the connections (Tarr,[11] p. 17). The NYC system of tide gates, which are described in detail below under flood control measures, is vulnerable to malfunctions under certain flooding conditions. Tide gates will open to release combined sewer system flows only where a half foot head (differential between the height of the water in the sewer system and the water level) exists. Should the surface water level rise higher than the water in the sewers, the gates would not open. If tide gates were totally submerged for long periods of time, water would eventually back up on the streets.[g]

The sanitary sewer system itself would probably not be prone to flooding directly. The top of a sewer is typically located about six feet below ground, and the bottom can extend deeper than ten feet depending on the diameter of the pipe. Even if these were located below sea level, flooding would only impact them through breakage and infiltration. If soils were to become sufficiently waterlogged, freezing during winter months could increase the frequency of breakages.

---

[g] The sections on the sewer systems was based on information provided by Robert Smith and Peter Young of Hazen and Sawyer Engineers, Inc., New York, NY.

Hazardous Materials Infrastructure

Pipelines are a major vehicle for the transport of hazardous materials. To the extent that they are vulnerable to breakage, the metropolitan area is vulnerable to the release of hazardous materials. In the U.S., 1.7 million miles of pipeline carrying hazardous materials and gas are under federal jurisdiction (U.S. Department of Transportation,[12] p. A-4). Susceptibility to damage can occur as a result of temperature changes brought about by water surrounding the pipes and in soil, erosion of soil and other supporting material around the pipes, elimination of bridge crossings which have been the major vehicle of support, and hitting of exposed pipe by debris.

In addition to pipelines, infrastructure for other hazardous materials is vulnerable during flooding. Ironically, it is the release of hazardous materials that often accompanies natural disasters that can produce most of the reported injuries and illnesses. According to the U.S. Geological Survey (USGS), during the 1989 earthquake in Loma Prieta, CA, one quarter of the injuries and illnesses that occurred after shaking and were not due to structural collapse were caused by toxic material exposure (U.S. Geological Survey,[13] p. A-17). These occurred among workers engaged in evacuation activities where releases of unknown substances from damaged structures occurred. A second source of worker exposure occurred during cleanup operations.

There are 52 inactive landfill sites in NYC, for which the New York State Department of Sanitation says evidence of illegal dumping of hazardous wastes exists (NYC Department of Sanitation,[8] p. 3-21). Many of these sites occupy waterfront locations, and migration of the wastes is more likely under higher water level conditions.

Roadways are common routes of hazardous material transport either as raw materials, intermediate or final products, or wastes. New York area roadways are particularly vulnerable to spillage owing to the large amount of economic activity involving these materials and the unique roadway configurations (Zimmerman and Gerrard[14]). Flooding impacts on the structural integrity of these roadways as discussed earlier, and hence, upon the risks associated with hazardous material spills.

Energy Production and Transmission Facilities

Energy production would be indirectly affected by high water levels and flooding by increasing the difficulty of obtaining fuel for energy production. High water levels flood energy production facilities and also impede goods movement where transportation access is blocked by water or land for the delivery of fuel oil or coal. This has already been a severe threat in some of the major floods that the U.S. has experienced in recent years.[h]

Pipelines are a common part of the energy transport system as they are of the waterworks system. If pipelines were to be permanently located in wet soils, they could fail under the pressure created by the freezing of waters surrounding them. This is already one mode of water and sewer pipeline failure in the metropolitan area.

---

[h] For example, during the Mississippi Floods of 1993, large portions of the Mississippi River were closed to barge movement for many days.[15] (See also U.S. Department of Transportation report on Midwest flood.[34])

What It Would Take to Prevent Damages

Ironically, the time period for which estimates of sea level rise have been made, i.e., 35 to 100 years, falls within many of the desired replacement or repair cycles for NYC infrastructure. Examples are 25–50 years for street repavement and 100 years for a water main (Wagner,[7] p. 28). Some accommodation to mitigation of the effects of sea level rise and storm surges could be made within these cycles.

*Modification of Materials and Structural Design:
The Case of Transportation*

Design and material standards for bridges and highways would have to be overhauled in order to cope with rising sea levels and to cope with extreme effects of temperature increases resulting from global warming. Steel structures imbedded in roadways and bridge decks would have to receive water protective coatings and in the long term be designed in such a way that the steel was more deeply imbedded in the structures. Expansion joints in roadbeds would have to be made wider to take the stress of increased temperatures. Maintenance would be critical to keeping the joints operating effectively to withstand pavement buckling. Considerable modification of existing bridges would have to be undertaken to bring them up to these new standards. Both New Jersey and New York have massive bridge reconstruction programs underway anyway in response to the National Bridge Rehabilitation Program. This provides an opportunity to prepare in advance for the needs projected to be created by global warming. The more dramatic approach to the problem would be to elevate roadways above the flood plains entirely. The use of flexible docking facilities for passenger services would have to be expanded in port facilities. Freight services would have to exclusively use lift-on/lift-off facilities.

Some existing construction projects serve as models for reconstruction. The North Channel bridge linking Cross Bay Boulevard in Queens and 165th Street in Howard Beach across Jamaica Bay was recently reconstructed. The previous bridge was entirely demolished, the current bridge had to stay in operation while the new one was under construction, and special structures were needed to withstand the tidal marine environment. The cost of the reconstruction, completed between 1990 and 1994 was $62 million (Parsons Brinckerhoff Quade & Douglas,[16] p. 1).

In addition to structural modifications, materials currently in use for roadway and rail construction will have to be modified to withstand heat, increased salinity, and increased exposure to water. These do not necessarily require new materials technology, but a more widespread use or transference of technologies usually used in rare situations.

Relocation and Redesign of Support Facilities

Pumps

Protective structures and water removal systems would have to be employed to reduce the consequences of water inundation. The number of pumps that many of the mass transit systems already use to remove water would probably have to be increased.

Pumping systems are currently one solution to flooding problems and could continue to perform that function in the future with added capacity. According to information from the Port Authority, the PATH system already has pumps operating 24 hours a day to remove normal seepage. Backup pumps are provided by sister facilities, such as the Holland Tunnel and the Jersey City Fire Department. The Jersey City Fire Department was particularly helpful in removing water that had flooded the PATH system using firefighting equipment. In general, fire departments provide considerable assistance in water emergencies because of their equipment and service capacity.[i]

New York City Transit, an entity within the Metropolitan Transportation Authority (MTA), also operates a large number of pumps to keep the subway tunnels free of water. In spite of the fact that much of the system is above sea level, the subway system is frequently inundated by water from a number of different sources not directly related to flooding, such as underground streams and high water tables in Brooklyn, water main and sewer breaks, and runoff and percolation from rain storms. In some parts of NYC the ground water is already brackish. As a result of the large amount of water entering the system, the MTA estimates that some 15 million gallons of water per day are pumped out. The MTA maintains 343 pumping stations for that purpose.

Monitoring Systems

Navigational aids and monitoring equipment will have to be designed, constructed and located so they will not be vulnerable to inundation.

Electrical Equipment

The location of energy production systems is a limiting factor in the protection of many types of infrastructure from flooding damage. A system may not be directly flooded, but it may suffer power malfunctions and outages if the energy production systems are flooded. Transportation systems, such as the PATH and NYC transit service between Brooklyn and Manhattan suffer outages when flooding of electrical facilities occurs. The PATH system maintains redundant diesel-fuel backup systems not connected to the major electrical utilities, and has elevation criteria for the location of electrical systems.

One key factor is how the electrical switch systems are positioned. Elevation on walls is more protective. Tide gate and regulator assemblies, which currently protect many areas from flooding, rely heavily on electrical systems. Regulator assemblies include one chamber which houses an electric motor and standby energy sources. A motor-operated system, rather than a mechanical one, is necessitated by the large volume of sewage that is handled. NYC is attempting to relocate the electrical systems, because of the corrosive atmosphere in the combined sewers (American Public Works Association,[17] pp. 42, 43, 47). This will also protect the systems against sea level rises.

Locational Changes: Reduction in Water Dependency

Much of the infrastructure that is currently water-dependent could be moved inland. For example, much of the metropolitan area's wastewater collection and

---

[i] Interview with Martha Gulick, Port Authority, New York, 1994.

treatment systems could conceivably have been located inland with flexible connections for wastewater discharges. The major obstacle to introducing such systems is the already large in-place fixed infrastructure including fixed distribution systems.

Flood Control Structures

To avoid massive relocation out of the flooded areas, an extensive system of flood control structures will be needed. Although a 4 foot rise in sea level would by no means exceed either the flood elevations or the flow rate or discharge of some of the worst flooding, the duration of flooding would be longer.

Flood control structures are a common and controversial approach to flood protection. In the worst floods that the U.S. has experienced, flood control structures have had a mixed performance record. They can increase the very flood elevations that they are designed to mitigate.

Tide Gates[j]

NYC and its surrounding metropolitan area already have a modest flood control system in place. Urban storm drainage systems handle flood waters on an intermittent basis. Tide gates are a major form of flood control for the area, and are connected to the combined sewer system. Tide gates prevent the backflow of water into sewer systems during high tides or flood stages. They consist of doors over the mouth of the sewers which are hinged at the top. They allow flow out through the pipe, and swing shut when the force of water is against them, *i.e.*, attempting to enter the pipe. In New York City and the metropolitan area, tide gates are used in conjunction with combined sewers, but not sanitary sewer systems. Combined sewer flows are currently directed to waterways, whereas sanitary sewer system flows are diverted to sewage treatment plants. NYC is currently embarking upon a plan for containment and treatment of combined sewer overflows.

All combined sewers in Manhattan up to the North River Sewage Treatment Plant (about 137th Street) require tide gates. North of that point and in most parts of the Bronx, tide gates are not needed because of higher elevations. The rest of NYC requires tide gates. The current inventory of tide gate/regulator combinations in the metropolitan area is approximately as follows:

New York City	500
Northeastern New Jersey	200
Yonkers	10
Long Island	None

Although discharge through regulators can occur under submerged or unsubmerged discharge pipes, tide gates are designed to be unsubmerged most of the time. When they are submerged, they allow no discharge through the pipe, which results in flooding of the streets. To completely reconstruct the tide gates would cost a few hundred thousand dollars per structure. In spite of existing protective

---

[j] Information on tide gates was provided in part by Robert Smith and Peter Young of Hazen and Sawyer Engineers, Inc., New York, NY.

measures, the city and the metropolitan area experience widespread, frequent and occasionally devastating flooding. This is reflected in the very large number of flood insurance claims that are filed under the National Flood Insurance Program in the metropolitan area each year. The city and the metropolitan area, in fact, are near the top of the list nationally in this respect, largely because of the number of people located in flood prone areas.

The states of New Jersey and New York ranked fourth and eighth respectively in the country in the flood claims paid between 1978 and 1987. Two areas within the metropolitan area, New York City and the Township of Wayne, New Jersey, ranked 7th (with 1,256 losses) and 9th (with 742 losses) respectively in the U.S. in the number of repetitive losses due to flooding between January 1980 and December 1989 (National Hazards Research and Applications Information Center,[18] pp. 41–42).

Flood-retarding Structures

Standard approaches to flood retardation are land reclamation, sea walls, flood gates, and breakwaters. The metropolitan area has had a long history of land reclamation. Sea walls are common along the Gulf and Atlantic coasts, protecting major cities. Bird cites costs of $500 billion dollars to protect the entire U.S. coastline analogous to the coasts of protecting the Dutch coastline (Bird,[2] p. 135). The Thames Barrier is one of the most famous structures protecting a dense urban area from flooding due to subsidence. It was completed in 1983. Similar structures exist in Tokyo and in the Netherlands. Floodgates cost an estimated $300,000 per kilometer to build and maintain. Overall the estimates cited by Bird on a per unit basis are as follows:

> It has been estimated that the raising and elaboration of coastal defenses to match a sea level rise of 20 cm along about 250 km of coastline would cost about $1 billion; for a 1-m rise the cost would be about $10 million per kilometer (Bird,[2] p. 136 citing, Goemans[19]).

The construction of relatively massive levees for urban areas built to Corps standards is estimated at about $1 million per mile. Costs are highly variable, varying according to the availability of the raw materials, the nature of the bed upon which it is built, etc.

Earlier estimates of construction costs at a time when many of these structures were built are shown in TABLE 4. Each of these facilities requires considerable

TABLE 4. Earlier Estimates of Construction Costs of Waterfront Structures

	1971 Dollars	1993 Dollars[a]	Assumptions
Seawalls	$1640/m ($500/ft.)	$5,439/m ($1,643/ft.)	Assumes a location of raw materials far from sources
Bulkhead	$245/m ($75/ft.)	$820/m ($251/ft.)	Assumes a low structure
Groins	$300–1000/m ($100–300/ft.)	$1,004–$3,347/m ($335–$1,004/ft.)	

[a] Earlier figures were converted to 1993 dollars using the fixed-weighted price index for gross domestic product, gross private domestic investment: fixed investment for nonresidential structures (Council of Economic Advisors, *Economic Report of the President, February 1994*, p. 274.) Original figures in 1971 were obtained by Yasso and Hartman[4] from the U.S. Corps of Engineers shore protection guidelines and are approximate only.

investments in maintenance and capital construction. In some cases, maintenance is to prevent undermining of the structure, in other cases to repair breaches.

Pumping and Drainage

Extensive and continuous use of pumping and drainage systems would be required to accompany these structures at least as an interim measure to cope with flooding. Although there is no technological limitation to employing such a measure, in the long term it could increase energy demands, which would only exacerbate global warming.

Synopsis

Scenario 1 suggests substantial in-place changes in infrastructure to resist flooding from sea level rises. It confronts a very large in-place system, which is difficult to modify without explicitly integrating the unique flood protection needs posed by sea level rise into ongoing capital improvement and planning programs. Examples of just some of the changes that would be required if a purely reactive option were followed are as follows:

- Ports for passenger service would have to incorporate more flexible docks. Roll-on/roll-off facilities for freight might become untenable.
- Most tide gates as currently constructed and situated would become completely submerged and would need to be relocated or reconstructed.
- The vulnerable points of transit tunnels—the shafts, vents, approaches—would have to be elevated. Transit systems would require continual pump operations to remove water. An upscaling in the use of pumps, however, would add to the greenhouse gas burden in the long term.
- Construction of a considerable number of sea walls would be required where infrastructure could not be retrofitted or relocated.

SCENARIO 2: LEVERAGING INFRASTRUCTURE AND LAND USE TO RESPOND TO AND RETARD GLOBAL WARMING

It may not be possible to implement *Scenario 1* because of the expense and engineering infeasibility due to intractable water dependency and large in-place investments for certain kinds of infrastructure. In addition, in the long run, these preventive measures may have more of an adverse impact on global warming, because of increased energy use associated with protective infrastructure.

Another option presents itself: that of massive relocation of both infrastructure and land use. Although this is perhaps the most radical solution, it presents the opportunity for altering land use patterns and activities in a direction that could reduce global warming in the long term.

Recognizing that a lot of changes in land use will occur anyway, regardless of what happens to infrastructure and regardless of global warming, a key issue is whether there is some increment that could be changed in the future that could result in a reduction in global warming. Some of this change could be guided by infrastructure, since infrastructure is famous as a mechanism for guiding land use.[k]

Land Use Activities to Promote Reductions in Global Warming

There is a considerable debate about the nature and feasibility of arriving at land use measures that would reduce global warming. This is compounded by the difficulty in linking land use measures to reductions in combustion emissions from both stationary and mobile sources. The U.S. Office of Technology Assessment (OTA)[20] attributes the bulk of greenhouse gas emissions to just three factors: population growth, miles traveled per person, and the emissions of greenhouse gases per unit of travel. The latter two factors can either be approached by changing transportation patterns and/or changing land use patterns that determine many of these transportation patterns.

Two approaches are set forth here that aim at reductions or improved efficiency in combustion processes for stationary and mobile sources respectively:

- Energy-conserving/emission-reducing means for space heating
- Energy-conserving/emission-reducing modes of transportation

Recognizably, uncertainties and debates about their effect and feasibility exist. The two approaches are briefly discussed first in general terms and then in terms of their relevance to the metropolitan area.

Energy-conserving/Emission-reducing Means for Space Heating[l]

Space heating accounts for a large percentage of energy consumed and emissions produced from stationary sources. For example, in New York State, 62% of the energy consumed in residential units is for space heating and the comparable figure for commercial units is 58%.

Higher Density Development

Given a constant population, there are a number of different ways of promoting less energy use per capita, primarily for space heating. In terms of land use changes, one approach is to promote higher density development. All other things being equal, high-rise development tends to be more energy efficient on a per household basis. For example, according to the Energy Information Administration,[21] electricity consumption for space heating in the northeast is 21.9 million Btu per household (4,423 kWh) for single-family housing and 10.1 million Btu per household (2,972 kWh) for multifamily housing (2 or more units). Unit energy

---

[k] At the height of the growth control movement of the 1970s, for example, book and report titles like *Land Use and the Pipe* and *The Growth Shapers* characterized the role of infrastructure in shaping land use patterns.

[l] Data sources were provided by Harry Sheevers of the Division of Policy Analysis and Planning of the NYS Energy Office.

TABLE 5. Unit Energy Consumption (UEC) for Primary Heat in Residential Units by Type of Residence, New York State, 1993

Service District	Single Family	Small Multifamily	Large Multifamily	Mobile Home
	(in kilowatt-hours)			
Consolidated Edison	3,266.8	4,342.7	2,074.1	
Long Island Lighting	8,476.9	5,484.3	2,274.9	7,901.2
Orange & Rockland	9,427.6	5,729.7	3,360.2	

NOTE: UEC figures given are for conventional primary heat only under normal weather conditions. *Small Multifamily*: 2–4 family houses, condos/coops similar to 2–4 family houses, and cases where no residence type was specified. *Large Multifamily*: Includes apartment buildings.
SOURCE: Regional Economic Research, Inc., "Residential End-Use Data Analysis."[31] New York: Empire State Electric Energy Research Corp., January 1994.

consumption figures, shown in TABLE 5 for selected utility districts in the New York State portion of the metropolitan area, are clearly lower for large multifamily structures. On a square footage basis, the patterns are less clear. Savings due to the insulating effect of a high rise structure may be offset by large nonresidential areas within such buildings that require heating. Moreover, heating depends on type of construction, age of structure, configuration, type of insulation and other factors, which make estimates highly variable.[m] An important contribution of high density structures is that they potentially reduce emissions from transportation facilities by minimizing overall travel and providing the density to make mass transit viable.

The metropolitan area varies dramatically in the prevalence and distribution of high density development. Although 40% of land use area in New York City is devoted to residential development (see TABLE 6), for example, multifamily residential buildings (exclusive of those in multiple-use structures) account for a relatively small proportion of the city's land use. Staten Island and Queens have relatively few multifamily structures: 1% and 9% of the land use respectively. Brooklyn, The Bronx, and Manhattan have somewhat larger percentages: 14%, 14%, and 27%, respectively. Thus, there is considerable room for higher density with the very considerable caveat, however, that community character and cohesiveness might be altered substantially to accommodate such increased density.

The choice to increase densities in certain areas of the city if it were found to reduce emissions of greenhouse gases both directly and from transportation would require a different configuration of infrastructure facilities and in some cases different kinds of infrastructure. Were infrastructure to be redesigned or relocated to guide redevelopment to encourage such higher densities, it could be done in the following manner:

- By restricting access to water supply and wastewater distribution points to certain discrete locations. The location of the lines themselves are difficult to restrict, since they are already in place.

---

[m] The comparable figures for energy consumption per unit floor area for different size housing is 0.64 kWh per 1000 square feet for single family housing units and 0.90 for multifamily.

TABLE 6. Distribution of Land Uses: Waterfront vs. Citywide Land, New York City

	Waterfront[a] (1990)	All NYC[a] (1994)
Federal, State and City parkland (includes natural or undeveloped land, active recreation areas, land abutting transportation corridors)	42%	35%
Industrial	31[b]	4
Residential	20[c]	40
Commercial	6[d]	3
Transportation (including parking), utilities		10
Public Facilities and Institutions		8
Undeveloped (Vacant)		8

[a] Of the 35% of total city land that is parkland, 13% is vacant; some of the land tabulated as residential is mixed commercial/residential use.

[b] This represents land zoned for manufacturing.

[c] This represents land zoned primarily for low density residential.

[d] This represents land zoned primarily C3.

SOURCES: The percentage distribution of land use and zoned uses for the waterfront is as identified in the New York City Waterfront Plan,[6] for 578 miles of shoreline, p. 7. The land use distribution for the City as a whole was obtained from the files of the NYC Department of City Planning, dated 10/11/94. Citywide data are based on lot area, and some lots whose dimensions were not known are not included.

- By increasing park land and open space around dense structures to discourage the use of such space for individual housing units.
- By providing better transit links: this is the key. Many high-rise developments do not provide for mass transit links.

Rethinking Waterfront Development

According to the U.S. Census, 53 percent of the population of the United States lives in counties that are within 50 miles of a coastal shoreline. This percentage has remained about constant at least since 1960 when these figures were tabulated. The area amounts to 467,000 square miles. The Atlantic portion of the population figure accounts for about a quarter of the total, and this percentage has also remained constant since 1960.[n]

New York State has the highest population density per shoreline mile of any coastal state, and New Jersey ranks third (after California) (Gornitz, White and Cushman,[22] pp. 2364–2365). New York State has an estimated 6,738 people per shoreline mile and New Jersey has 3,898.

This high level of density in the coastal areas of the metropolitan area has locked the area into certain infrastructure patterns, several of them unique. For example, NYC has some of the largest sewage treatment plants in the country and probably the world, all of which have waterfront locations.

[n] U.S. Census, *Statistical Abstracts of the U.S.—1992*. Page 28, Table 32.

In the short term, a key response option is to move development as well as infrastructure inland. This would be one way of protecting land uses from coastal inundation. Waterfront location does not necessarily mean water dependency. Some infrastructure facilities could move from the waterfront under a more radical scenario.

In order to understand the ramifications of altering the land use balance between shoreline areas and inland, it is useful to look at the land use patterns in each of these areas. Land use patterns in waterfront areas and elsewhere in the metropolitan area differ substantially from one another. For example, the distribution of land use along the city's waterfront and in New York City as a whole is shown in TABLE 6. Waterfront uses tend to be predominantly devoted to industrial uses (often abandoned) and open space. In contrast, land use in the city as a whole tends to be relatively more residential and far less industrial, with open space amounting to about the same percentage of the land area.

In the metropolitan area, growth from new development on the waterfront or redevelopment is continuing. For the NYC portion of the metropolitan area, some of these developments were listed in the Waterfront Plan as existing redevelopment projects and proposed redevelopment opportunities. These are shown in TABLE 7.

Were waterfront development to continue as it has been rather than locating inland, conditions to the approval of waterfront projects could incorporate both design features for flood protection as well as energy conservation to contribute to the reduction of the global warming effect altogether. The flood protection conditions would not only serve the purposes of protection against sea level rise but would also protect against normal flooding. The energy conservation conditions would not only address global warming, but air pollution as well. This is by no means a novel situation. Many of these energy conservation conditions are already in place, and probably do not introduce much that is new over and above what energy and air quality plans are now already addressing.

Energy Conserving/Emission Reducing Modes of Transportation

In the U.S. as a whole, transportation accounts for about 40 percent of total net energy consumption and about a quarter to a third of the gross energy con-

TABLE 7. Existing and Proposed Redevelopment Areas Within the Designated New York City Waterfront

	Existing Number of		Proposed[a] Number of	
	Sites	Acres	Sites	Acres
Bronx	2	70	6	40
Brooklyn	3	48	13	243
Manhattan	2	41	13	118
Queens	3	380	7	271
Staten Island	3	84	5	41

[a] Only includes acreage for sites where acreage of project was indicated in the NYC Comprehensive Waterfront Plan. Some of these areas as of 1994 may have already been developed, and others may have been abandoned.
SOURCE: Tabulated from the NYC Department of City Planning, NYC Comprehensive Waterfront Plan (1992),[6] Tables 6.0, 6.1, 6.2, 6.3 and 6.4.

TABLE 8. Relative Energy Intensity or Utilization of Selected Alternative Modes of Passenger Transportation

	Oak Ridge[32] (1994) 1991		Gordon[33] (1993) 1989	
	A. Btu per passenger-mile			
Mode	Load/ Occupancy per vehicle	Btu	Load/ Occupancy per vehicle	Btu
Automobiles	1.6	3,604	1.7	6,530–8,333
Buses				
Transit	9.7	3,811	10.2	3,761
Intercity	23.2	962	21.5	939
School	19.4	848	—	—
Rail				
Intercity	20.1	2,503	20.5	2,537
Transit	20.2	3,710	22.8	3,534
Commuter	34.0	2,993	36.1	3,138
	B. Btu per vehicle-mile			
Mode	Btu		Btu	
Automobiles	5,767		6,530	
Buses				
Transit	36,939		38,557	
Intercity	22,310		20,176	
School	16,419		—	
Rail				
Intercity	50,321		52,107	
Transit	74,864		80,550	
Commuter	101,843		113,228	

SOURCES: Oak Ridge figures are from Stacy C. Davis, Transportation Energy Data Book: Edition 14, May 1994.[32] Deborah Gordon, Steering a New Course: Transportation, Energy, and the Environment.[33]

sumed[o] OTA estimates that the contribution of transportation to carbon emissions is 32%, which is about equal to the share of energy consumed (U.S. Congress, Office of Technology Assessment,[23] p. 169). Measures of energy use per passenger are extremely sensitive to the load factor or occupancy that is assumed, and have to be interpreted in that light. The relative use of energy on a per capita basis for each mode is shown in TABLE 8 from two different sources of data. Energy intensity per passenger-mile for mass transit is about comparable now to that of automobile travel, primarily because of increased fuel economy in autos but also

---

[o] NYS Energy Office, Vol. 3, 1994, p. 18, 177. According to the U.S. DOT and the U.S. DOE, transportation accounts for only about a quarter of the total gross energy consumption, and this percentage has risen very slowly from 23% in the mid-1950s to the current level of 27% (U.S. DOT,[12] 1993, p. 153). The difference between the two figures is attributable to the fact that New York State figures are net figures, i.e., independent of the contribution of energy use at power plants. Federal figures, in contrast, include that energy, making the transportation component relatively smaller because it includes less electricity from power plants.

because ridership in transit has reduced the load making the intensity figures seem higher. If ridership were at capacity per passenger levels of energy consumption would be more favorable for transit as the more energy efficient mode.

Although energy efficiency has improved over the past couple of decades for automobiles, the amount of travel has increased to an extent that offsets fuel savings, and thus, total amount of energy consumed by automobiles continues to rise. In the past couple of decades, the rate of automobile use has increased over previous years and faster than the rate at which population has grown (U.S. Department of Transportation,[12] p. 195).

Current approaches toward energy-efficient modes of transportation are the use of alternative fuels, fuel economy, and more energy efficient or cleaner (less emission-producing) modes of travel such as mass transit and bicycling. The latter approach implies at least some marginal changes in the pattern of land development. Changing settlement patterns was proposed as one policy option in the OTA study on global warming (U.S. Congress, Office of Technology Assessment,[23] p. 169). Recognizing that during a period of slower population growth little in the way of major changes can be expected, OTA noted that:

Emissions can be reduced in the long run by changing patterns of settlement to reduce the need for travel or to increase the utility of mass transit. This can be accomplished through higher densities or through mixing uses so that residences, jobs and services are roughly balanced at a local scale. When more destinations are close to home, more trips can be made by foot, and public transit can serve more trips effectively.

Thus, to the extent that infrastructure can shape future land use at least at the margin, minimizing energy utilization by transportation argues for a multimodal approach. Long distance travel and circumferential transportation in less dense areas would rely on bus and auto travel, and inner city travel would primarily be via bus and mass transit, with parking facilities available at the periphery. Bicycling would be consistent with minimizing energy utilization regardless of location. This urban design dimension to environmental problems associated with transportation is not new. It was proposed, for example, by Bleviss and Walzer[24] (p. 104) in an article addressing the energy and environmental problems of automobiles.

A considerable amount of adaptation of the transportation system would be needed to promote such a concept, however. Mass transit in the NYC metropolitan area as in most cities, aims at promoting suburban/central business district travel, the so-called "hub-bound" travel, but is deficient in providing circumferential connections. Projections tend to point to more non–work-related automobile trips in the future, for recreation, home to school travel, and shopping. If this is realized, then either rail or bus transit links would have to be provided to promote circumferential travel via mass transit.

Infrastructure as a Major Energy User

Infrastructure can be a major energy user and contributor to the greenhouse problem. Addressing this problem alone could be a major contribution of the metropolitan area to the reduction in global warming. In terms of setting priorities, however, it is unclear how this compares to other forms of energy use because of uncertainties in the data on energy use and the highly decentralized and dispersed nature of these facilities.

Infrastructure uses energy for lighting, temperature control, and equipment and processes necessary for the provision of infrastructure services. Linkages between design and energy use exist for infrastructure as they do in all other sectors.

Examining energy use for two major forms of infrastructure services—transportation and wastewater treatment—may shed some light on this, for example:

Wastewater Treatment. The NYS Energy plan estimates that 570 wastewater treatment plants in the State consume 1.5 billion kilowatt hours (kWh) of electricity per year to treat sewage and manage the sludge residuals. The Plan estimates that 75% of these plants are in the metropolitan area including Long Island, accounting for 60 percent of the wastewater flow treated in the State. Some of the energy savings suggested in the NYS Energy Plan are aimed at how the plants provide oxygen to the systems. The Plan also points to energy production opportunities at the plants with alterations in aeration systems potentially saving between a third to over half of the energy demand from wastewater treatment in the State (New York State Energy Office,[25] vol. 3, pp. 488–489).

Transportation. As previously noted, transportation currently accounts for about 40 percent of the energy consumed in the U.S. and the same percentage in New York State (New York State Energy Office,[25] vol. 3, p. 18). A 1994 New York Bar Association conference report suggested some of the ways in which major transportation service providers as distinct from the users could reduce energy consumption:

> . . . state agencies such as the New York State Thruway Authority, the Triborough Bridge and Tunnel Authority, and the Port Authority could limit the unnecessary consumption of outdoor lighting along thruways and bridges, by utilizing energy-efficient bulbs and by replacing only a percentage of the lights now in use when the bulbs burn out (Ginsberg[26]).

CONCLUSIONS: RECOMMENDATIONS FOR POLICY, MANAGEMENT AND PLANNING

Altering existing infrastructure on an *ad hoc* basis in response to the flooding produced by rising sea levels would involve massive reconstruction and relocation. Some infrastructure could not be relocated because of its water dependency. A more fruitful approach is to integrate the process of change into ongoing capital improvement and planning programs within the metropolitan area and to link infrastructure changes to land use configurations that reduce greenhouse gas emissions.

In spite of the long replacement cycles for the NYC metropolitan area's infrastructure, a number of planning and capital improvement programs are underway to replace aging infrastructure into which solutions could be incorporated that are advantageous in preventing impacts of and reducing global warming in the metropolitan area. Programs to address aging infrastructure problems of the metropolitan area provide an opportunity to address global warming. The long-term cycles of replacement and repair would have to incorporate design and locational criteria to adapt to sea level rise. Institutions have to start now given the slow course of construction and the very long replacement cycles, many of which exceed the time period during which sea level rise will be felt.

The solutions addressing infrastructure and land use to reduce global warming as well as reduce or mitigate its consequences lie in using a broader planning context to address global warming in general and the NYC metropolitan area's contribution to it. With respect to infrastructure and land use, three planning processes are already in place which are consistent with global warming mitigation. These are State Implementation Plans for the Clean Air Act, State energy plans, and the Statewide Master Plans for transportation being prepared under the Intermodal Surface Transportation Efficiency Act of 1991 (ISTEA). Examples of mechanisms that pertain specifically to infrastructure aim at decreasing automobile emissions by means of declines in automobile travel or conversion to cleaner fuels and increased use of mass transit.

Institutional Consequences

Addressing global warming by means of ongoing programs has enormous institutional consequences that pertain to jurisdictional issues as well as information management (National Research Council[27]). Some examples are noteworthy.

Solutions to global warming require data bases that can address the questions posed by global warming. Data linkages have to be made that will enable planners to map predicted locations of newly flooded areas (and the more difficult to predict areas of erosion and accretion), and then to superimpose this information over land use types. Geographic Information Systems certainly provide the technical capability, but data base design is far from being able to provide the inputs to such systems. Inventories of existing infrastructure and flood elevations, which are a critical component of a vulnerability analysis for sea level rise, do not exist in a comprehensive way for the NYC metropolitan area. The USACE inventory of transportation systems is an important start in this direction. Without such information, we are unable to identify which portion of the infrastructure is susceptible to flooding.

Some programs that are in place to address flooding require changes to address global warming specifically. The National Flood Insurance Program (NFIP), for example, currently does not acknowledge sea level rise in its flood insurance policies (U.S. Office of Technology Assessment[28]). Maps do not indicate projected areas of sea level rise, nor is information on the maps sufficient to add this information easily (*i.e.*, no elevation contours are given on the maps). Relocation out of the flood plain is not directly encouraged by the NFIP, and has to be supported under separate programs with special, usually small budgetary allotments on an episodic basis.

In the next few decades, NYC and the metropolitan area are expected to make massive investments in infrastructure, which have not been required to incorporate the threat of sea level rise in the development equation. Route 9A, for example, along the west side of Manhattan is likely to be under construction as sea level changes if predictions are borne out. An estimated $1.5 billion investment is expected to address combined sewer overflows in the city, which takes into consideration regional flooding but not long-term sea level rise. Aside from new construction there is an ongoing program of reconstruction in almost every area of infrastructure in NYC. Just as flooding is incorporated into the design of these facilities, so should sea level rise be incorporated. The waterfront plan identifies a future of rebuilding and beautification. Incorporating sea level rise scenarios would be easy now, but impossible later.

REFERENCES

1. MILLER, T. R. 1988. Impacts of global climate change on metropolitan Infrastructure. *In* Proceedings of the Second North American Conference on Preparing for Climate Change.: 366–376. Washington, D.C.: Climate Institute.
2. BIRD, E. C. F. 1993. Submerging Coasts. New York: Wiley.
3. INTERGOVERNMENTAL PANEL ON CLIMATE CHANGE. 1990. Climate change: The IPCC scientific assessment. New York: Press Syndicate of the University of Cambridge.
4. YASSO, W. E. & E. M. HARTMAN, JR. 1975. Beach Forms and Coastal Processes. MESA New York Bight Atlas, Monograph 11. Albany, NY: Sea Grant Institute. January.
5. ORTOLANO, L. 1988. Predicting impacts of infrastructure on land use. *In* Public Infrastructure Planning and Management. J. M. Stein, Ed.: 189–207. Newbury Park, CA: Sage.
6. NEW YORK CITY DEPARTMENT OF CITY PLANNING. 1992. New York City Comprehensive Waterfront Plan: Reclaiming the City's Edge. New York: New York City Department of City Planning. Summer.
7. WAGNER, R. F., JR. 1984. Infrastructure issues facing the City of New York. Ann. N. Y. Acad. Sci. **431:** 27–33.
8. NEW YORK CITY DEPARTMENT OF SANITATION. 1992. A Comprehensive Solid Waste Management Plan for New York City and Final Environmental Impact Assessment. New York, NY. August.
9. METROPOLITAN TRANSPORTATION AUTHORITY, NYC TRANSIT AUTHORITY. 1993. Facts & Figures Book. New York: Metropolitan Transportation Authority.
10. U.S. ARMY CORPS OF ENGINEERS, FEDERAL EMERGENCY MANAGEMENT AGENCY, NATIONAL WEATHER SERVICE, NY/NJ/CT STATE EMERGENCY MANAGEMENT. 1995. Metropolitan New York Hurricane Transportation Study. Interim Technical Report. U.S. Army Corps of Engineers.
11. TARR, J. A. 1984. The evolution of the urban infrastructure in the nineteenth and twentieth centuries. *In* Perspectives on Urban Infrastructure. R. Hanson, Ed.: 4–66. Washington, D.C.: National Academy Press.
12. U.S. DEPARTMENT OF TRANSPORTATION. 1993. National Transportation Statistics. Washington, DC: U.S. Government Printing Office.
13. U.S. GEOLOGICAL SURVEY. 1994. The Loma Prieta, California, Earthquake of October 17, 1989: Loss Estimation and Procedures. Washington, DC: U.S. Geological Survey.
14. ZIMMERMAN, R. & M. GERRARD. 1994. Hazardous substance emergencies in New York City. Disaster Management **(6)**3: 133–140.
15. ZIMMERMAN, R. 1994. After the deluge. The Sciences **(34)**4: 18–23.
16. PARSONS BRINCKERHOFF QUADE & DOUGLAS. 1994. Field Notes **(4)**3. May/June.
17. AMERICAN PUBLIC WORKS ASSOCIATION. 1970. Combined sewer regulatory overflow facilities. A manual of practice. Washington, DC: American Public Works Association. July.
18. NATURAL HAZARDS RESEARCH AND APPLICATIONS INFORMATION CENTER. 1992. Floodplain Management in the United States: An Assessment Report. Vol. 1: Summary. Prepared for the Federal Interagency Floodplain Management Task Force. Boulder, CO: Univeristy of Colorado.
19. GOEMANS, T. 1986. The sea also rises: The ongoing dialogue of the Dutch with the sea. *In* J. G. Titus, Ed. Effects of Changes in Stratospheric Ozone and Global Climate, Vol. 4. J. G. Titus, Ed.: 47–56. U.S. Environmental Protection Agency. Washington, DC:
20. U.S. CONGRESS OFFICE OF TECHNOLOGY ASSESSMENT. 1989. Washington, DC: U.S. Government Printing Office.

21. ENERGY INFORMATION ADMINISTRATION. 1993. Household Energy Consumption and Expenditures: 1990. Washington, DC: Energy Information Administration. February.
22. GORNITZ, V., T. W. WHITE & R. M. CUSHMAN. 1991. Vulnerability of the U.S. to future sea level rise. Coastal Zone '91. *In* Proceedings of 7th Symposium on Coastal & Ocean Management. American Society of Civil Engineers. Long Beach, CA. July 8–12.
23. U.S. CONGRESS, OFFICE OF TECHNOLOGY ASSESSMENT. 1991. Changing by Degrees: Steps to Reduce Greenhouse Gases. Washington, DC: U.S. Government Printing Office. February.
24. BLEVISS, DEBORAH L. & PETER WALZER. 1990. Energy for motor vehicles. Scientific American. September: 102–109.
25. NEW YORK STATE ENERGY OFFICE. 1994. Draft New York State Energy Plan. Albany, NY: New York State Energy Office. February.
26. GINSBERG, W. R. 1994. The threat of global climate change—what can New Yorkers do?: State and local strategies to reduce greenhouse gas emissions in New York State. Report of the Environmental Section of the New York State Bar Association. January.
27. NATIONAL RESEARCH COUNCIL. 1990. Policy Implications of Greenhouse Warming. Washington, DC: National Academy Press.
28. U.S. CONGRESS, OFFICE OF TECHNOLOGY ASSESSMENT. 1993. Preparing for an Uncertain Climate. 2 volumes. Washington, DC: U.S. Government Printing Office.
29. BROCCOLI, A. J. 1996. The greenhouse effect: The science base. Ann. N. Y. Acad. Sci. **790:** 19–27. This volume.
30. MINTZER, I. 1988. Living in a warmer world: Challenges for policy analysis and management. Journal of Public Policy Analysis and Management **(7)**3: 445–459.
31. REGIONAL ECONOMIC RESEARCH, INC. 1994. Residential end-use data analysis. New York: Empire State Electric Energy Research Corp. January.
32. DAVIS, S. C. 1994. Transportation Energy Data Book: Edition 14. Oak Ridge, TN: Oak Ridge National Laboratory. May.
33. GORDON, D. 1991. Steering a New Course: Transportation, Energy, and the Environment. Washington, DC: Island Press.
34. U.S. DEPARTMENT OF TRANSPORTATION, FEDERAL HIGHWAY ADMINISTRATION. 1993. Transportation Roads and Bridges: 1993. Midwest Flood Recovery Task Force Report. August 31.
35. AMERICAN PUBLIC TRANSIT ASSOCIATION. 1992. Transit Fact Book. Washington, D.C.: American Public Transit Association.
36. ENO FOUNDATION FOR TRANSPORTATION. 1993. Transportation in America. Westport, CT: Eno Foundation for Transportation.
37. AMERICAN PUBLIC WORKS ASSOCIATION. 1970. Combined Sewer Regulation and Management: A Manual of Practice. Washington, DC: American Public Works Association. July.
38. NEW YORK METROPOLITAN TRANSPORTATION COUNCIL. 1994. Transportation Improvement Program. September 12.

DISCUSSION OF THE PAPER

QUESTION: What can you say about the Route 9A plans?

RAE ZIMMERMAN: Route 9A, as many of you know, is basically a waterfront roadway. I am not sure whether the flood elevations in the proposal alternative really take into account some of these additional rises that could occur as a result of sea level rise. There are many transportation experts in the audience who might address that.

QUESTION: Do you know when the Corps of Engineers report will be available?

ZIMMERMAN: That I don't know. The data tables that I was able to get came out in June 1994. It was magnificent to see five agencies work together and get that data. They are working on the report. [The report became available in November 1995.]

QUESTION: Is the electrical power system included?

ZIMMERMAN: I didn't see electric power systems in the report. I believe that they only looked at transportation systems. Environmental facilities as well as other things should be looked at.

Impact of Global Warming on Water Resources

Implications for New York City and the New York Metropolitan Region

ROBERT ALPERN

Re-engineering and Strategic Planning Group
New York City Department of Environmental Protection
59-17 Junction Boulevard
Corona, New York 11368

Global warming effects—and the changes in atmospheric chemistry that produce global warming—may significantly change the demands placed on the water supply and wastewater systems of New York City and the New York Metropolitan Region and these changes may in turn significantly change the demands placed by those systems on regional land use and finances. Given the uncertainties—uncertainties about the nature, magnitude and timing of the changes—global warming sets a profound challenge for public policy: how to initiate and sustain a timely, non-wasteful "adaptive management" response.

THE SYSTEMS AT RISK

First, let's review the systems at risk.

New York City (NYC) and much of the surrounding region take their water supply from three major hydrologic systems (FIG. 1). The city's Delaware Watershed is at the headwaters of the Delaware River Basin, a hydrologic system including parts of New York State, New Jersey, Pennsylvania, and Delaware. The city's Catskill and Croton Watershed and its Hudson River Pumping Station (for emergency supply) are within the Hudson River basin. And the wellfield of the Jamaica Water Supply Company in southeast Queens draws on part of the groundwater system underlying all of Long Island.

New York City and much of the region discharge storm water and sanitary sewage—the used water supply—into the estuarine system that includes the lower Hudson, New York–New Jersey Harbor, Long Island Sound, and the near Atlantic Ocean (the New York Bight).

In each of these hydrologic systems, actions in one part affect all other parts, often with significant ecological and political consequences. And New York City in particular ties them all together. The effects of global warming only dramatize these profound interdependencies.

WATER SUPPLY POLICY

One result of the global warming and its effects—sea-level rise especially—could be greater regional dependence upon the NYC water supply. On Long

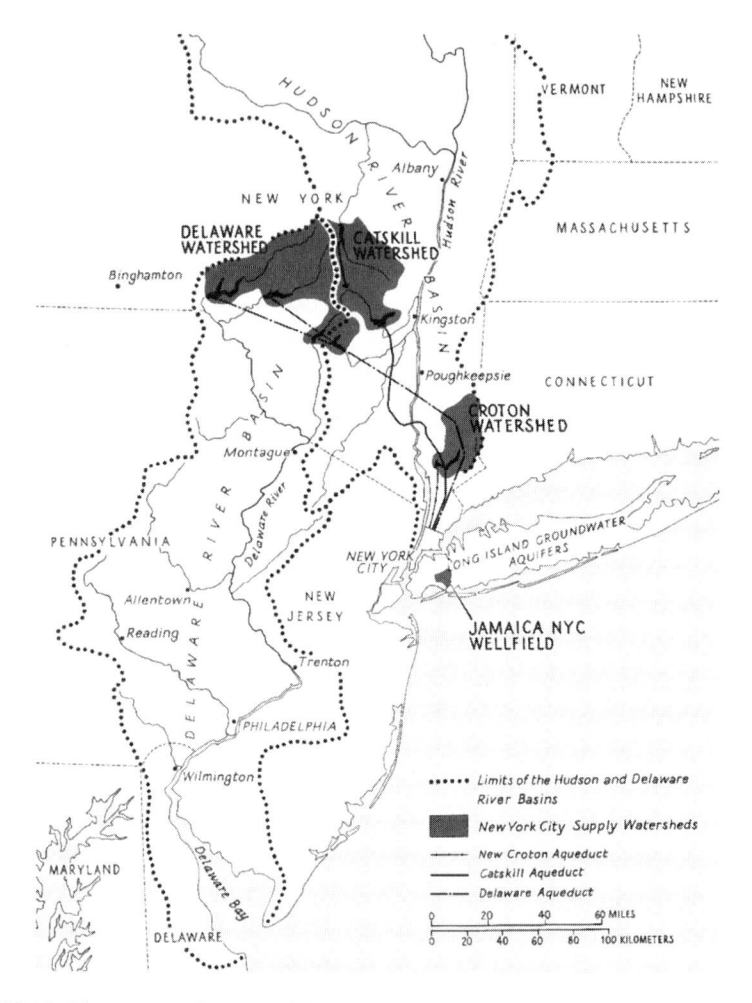

FIGURE 1. The sources of New York City Water. (Map courtesy of Citizens Union Foundation, New York, NY.)

Island, communities might turn to the NYC system if sea-level rise leads to significant salt water intrusion into the Long Island aquifers. In the Delaware River Basin, additional water releases might be required from NYC reservoirs—reservoirs that supply half the city's demand—if sea-level rise leads to a significant upriver shift in the Delaware River salt front (the shifting line where ocean water and fresh water meet), a shift that necessitates an augmentation of Delaware River flow in order to protect Philadelphia's water supply and the viability of Delaware Bay fisheries. In the Hudson River basin, many more communities might tie into the New York City system—a right many of them have had under state law since 1905—if precipitation becomes unreliable or an up-river shift in the Hudson River salt front knocks out the intakes for river-fed supplies like Poughkeepsie's. At the same time—with changes in the seasonal pattern of precipitation and possibly lower precipitation overall, with higher evaporation, and with demands driven by an increase in annual degree-days—NYC might have less water to share.

And New York City might have fewer opportunities to augment its supply, since the options for new and expanded sources may themselves be threatened by the impacts of sea-level rise. Thus, use of Brooklyn–Queens ground water could be limited or precluded by salt water intrusion from the ocean and the Sound, and use of intakes on the Hudson River could be limited or precluded by northward displacement of the Hudson River salt front.

POLICY FOR THE HUDSON ESTUARINE SYSTEM AND ITS COASTAL ZONE

Global warming and its causes and effects may also have significant implications for policy in the estuarine system—New York–New Jersey Harbor, Long Island Sound and the New York Bight—and its coastal zone.

The geometry of the water bodies and their shorelines, the frequency and intensity of storms, the direction, steadiness and intensity of winds, the temperature and chemistry of the water, and the location of the great ocean currents—all could change. Those changes, in turn, could have major impacts on biota: destroying and creating habitat; possibly encouraging the conditions that in the past have led to low- or no-oxygen (anoxic) events, to coastal wash-ups of floatables and debris, and to green, red and brown tides; transforming the abundance, health, and composition of aquatic and coastal plant and animal life.

It is assumptions about estuarine biota, of course, that are driving multibillion-dollar decisions under the federal Clean Water Act and Coastal Zone Management Act about wastewater treatment, nonpoint source control, and the dredging of sediments.

WHERE DECISIONS ARE BEING MADE

The water supply issues raised by global warming are being addressed by the New York City Department of Environmental Protection in two of the department's current long-range planning studies for water supply: the Hudson River Alternatives Study and the Brooklyn–Queens Aquifer Study. They are not being addressed, at least not in a coherent way, by official New York State policy. By state law, the state's official water resources policy is embodied in the Statewide and Substate Strategies approved by the New York State Water Resources Council, which for all practical purposes has ceased to function.

The issues being raised by global warming in the estuarine system are being addressed by the New York City Department of Environmental Protection in two of the department's planning studies for clean water: the Jamaica Bay Comprehensive Watershed Plan and the East River Upper Bay Water Quality Plan. But they are barely touched in current policy processes at the state or federal level, including the National Estuary Programs for the estuarine system: the Long Island Sound Study, the National Estuary Program for New York–New Jersey Harbor, and the Bight Restoration Program.

THE ECOSYSTEM PROTECTION APPROACH

Increasingly, federal and state environmental policies favor a "place-based" ecosystem protection approach:

- policy that is tailored to the realities of specific geographic areas;
- policy that integrates long-term environmental management with human needs: public health, economic renewal, and environmental justice;
- policy that strives for fairness and consensus among the "stakeholders" affected by and responsible for the area;
- policy that takes into account the relations among all environmental media: water, land and air; and
- policy that is based on sound science; priorities that are based on risk of harm; adaptive management that adjusts plans as information improves.

And so they should.

The ecosystem protection approach will be the context for future decisions affecting the water supply and estuarine systems; from "whole community" and "whole basin" planning in the NYC watershed to revisions in the National Estuary Studies for the Sound, the Harbor and the Bight.

The unstated assumption of the approach is stability in the natural context, as a fixed base for negotiation and trade-offs. But that assumption may not be tenable.

CONCLUSION

The greatest problem is uncertainty.

There seems little doubt that the apparent climate stability of the first half century of the twentieth century—the period that dominates our water supply and estuary records—was illusory. Investigation of past climate in this region, assisted by tree-ring analysis at the Lamont-Doherty Earth Observatory, has disclosed quite different patterns in the 18th and 19th centuries, much more variable and extreme.[1] Meanwhile, sea level has been rising in the Hudson and Delaware Rivers since at least the turn of the century, for reasons apparently unrelated to global warming.

The global warming scenarios simply make the uncertainties more unnerving and intolerable.

ADAPTIVE MANAGEMENT

Global warming is perhaps the most dramatic context for dealing with adaptive management. But, even well short of a clear signal that global warming has begun, adaptive management seems to me to be the emerging issue of the 1990s: how to negotiate environmental policy and policy implementation in a context of scientific uncertainty, technological change . . . and, perhaps changes in social preferences. So far as I can tell, the literature on this—including the legal literature—hardly exists.

Adaptive management, of course, is really a sub-set of the classical legal tension between reasonable expectation and a changing world.

It seems to me the critical issues are these:

- how to create agreements that permit adjustments, or require renegotiation, based on performance measurement and better information about the nature of the problem and the means to address it;
- how to establish triggers, mechanisms and conditions for adjustment; and
- how to establish triggers, mechanisms and conditions for renegotiation.

I urge you all to take up the challenge.

REFERENCE

1. BAILEY, B., K. WEBSTER & R. STEWART. 1985. Long Island Precipitation Patterns and Drought Probability, Atmospheric Sciences Research Center, State University of New York at Albany. March.

DISCUSSION OF THE PAPER

VICTOR WOUK: Sir, this general topic will be discussed this afternoon. Assuming that energy is no problem, what is the magnitude of energy required to help the water problem by desalinization?

ROBERT ALPERN: I don't have an answer for that, because I don't have the figures for desalinization. Rae, do you have any figures on that? Rae, did you develop figures on this in the past, because I think you did address this at one point?

RAE ZIMMERMAN: I have some older figures, but not at hand. It will be expensive.

ALPERN: We don't know what the technology in ten or fifteen years will be. Maybe it will be less expensive. Right now, on the water supply side, we have a relatively benign situation in terms of operating expenditures because so much of the system operates on gravity. We haven't had to worry about water quality issues. If in fact we have to go to filtration, for example, for the water supply, then we have major new operating expenditures, of which power—energy—expenditures are an important part. Rae did mention that there are significant power expenditures that relate to the sewage treatment plants right now. Even apart from the possible implications for desalinization—assuming that was going to be an acceptable technology to the consumer, a question I have raised in terms of the scenarios—there is no question that we are facing the possibility of major power increases for the system.

LARRY SWANSON: Bob, one thing that you didn't bring out in your excellent comments was the value of waste prevention or water conservation, as a means to get around some of the tremendous expenses associated with both use, in the delivery system, and the treatment after use. Since New York City is right at the top of the list in water consumption per capita, that might be a low-tech technology that you could implement to help resolve some of the problems you have addressed.

ALPERN: Yes, I think that in fact the city has probably one of the most aggressive water conservation programs in the country right now, including everything from toilet retrofits to examining our own facilities. There is some evidence that we have tightened our belt already. To my surprise, as somebody who focused on the water resources issue initially from the point of view of water supply, it really is the problem of the capacity of the sewage treatment plants that has driven the water conservation program. At some point, though, we are going to reach some critical minimum beyond which we cannot really impose on the public. There will be a revolt. Right now, we are basically 1.4 billion gallons a day, and I would guess that we have in sight now the possibility of going to 1.1 billion gallons a day, but lower than that I don't think we can go. Clearly, that is an important low-cost solution. Any kind of source reduction and water conservation is going to be important to us.

QUESTION: In addition to technological responses, are there any educational programs to encourage water conservation?

ALPERN: There are and not only for average citizens. There is an increasingly aggressive effort to engage the business community on ways of "greening," not only to make them environmentally good citizens, but also to make them competitive in what are going to be the increasingly green markets, the increasingly green operations of the future; the things that Regina talked about. Clearly, that is an important way to go and there are conservation programs, education programs, out there. The extent to which we can rely on soft programs of that kind is one of the issues. Since many of the people in the audience are engineers, you know that the predilection of the engineer to go with what is knowable, not the stuff that is problematical. This may be an important aspect of the adaptive management approach. I would caution, though, in terms of adaptive management, not to overrely on things that are problematical, because that way we may find ourselves in a situation where we have just avoided the problem. But there is no question that education is an important part of what has to be done.

QUESTION: Are you taking sufficient account of the fiscal realities? With the debt facing governments, it is difficult to see how major structural changes can be made.

ALPERN: It really hasn't been given as much attention. I had hoped when the Citizens Budget Commission did a major report on New York City's water utilities a year or so ago, that would come across loud and clear, but unfortunately it did not. The department is now trying to make this point within city government. One of the things that we can do that other agencies can't, because of the special institutional arrangements that we have, is to use additional head count and operations to set in place programs that in the end will reduce future capital expenditures. But the city fathers, and for that matter the fiscal oversights even in the public-interest sector, seem to be focused on head count reductions and operating expenditure reductions without understanding the tradeoffs with capital expenditures. Again, as far as I am concerned, the real issue for this conference is adaptive management with the financial picture as the major context for that adaptive management strategy.

QUESTION: Mr. Alpern, in view of the uncertainties that you recognize, I am wondering if a primary tool of adaptive management—where you essentially want to avoid the extremes of doing nothing or doing everything—should be the classical operations research. Is that embodied in the analyses?

ALPERN: Yes, the fact is that we are doing operations research. The reinvention part of my title is to a large degree a modernized version of classical operations research. The problem is that we are getting in a highly politicized context where the kind of rational approach that operations research represents has an upstream fight to make itself felt. The point about getting a more rational picture on the relationship of capital and expense budget for the Department of Environmental Protection is an example.

Metropolitan New York in the Greenhouse
Air Quality and Health Effects

L. I. KLEINMAN AND F. W. LIPFERT

Department of Applied Science
Brookhaven National Laboratory
Upton, New York 11973

INTRODUCTION

A variety of potential effects on human health resulting from climate change have been identified in several assessments. According to an international panel[1] they include direct effects of extreme temperatures on cardiovascular deaths, secondary effects due to vector-borne diseases or crop yields, and tertiary effects such as those that might arise from conflicts over freshwater supplies. To this list we add the secondary effects of increased air pollution, which may result either directly from climate change or indirectly from increased air conditioning loads and the corresponding pollutant emissions from electric utilities. Higher ozone concentrations have been linked to increased ambient temperatures by both theory and observations of monitoring data. A similar association with particulate matter has been limited to observations, thus far. The pollution-heat linkage has been recognized before,[2] but health effects have not been evaluated in terms of predictions of the joint effects of both agents.

This paper has been prepared in two sections. First, we discuss the ozone situation with special reference to the Northeast Corridor and New York. In the second section, we present estimates of the health effects of climate change on New York and discuss some mitigation options.

AIR QUALITY

New York City, like most large cities, has an air pollution problem. We tend to notice this problem most in the hazy, hot, and humid days of July and August. Haze can reduce atmospheric visibility from a value greater than 50 km on a crisp cold day to 5 km or less. Although this haze has some natural components, it consists mainly of pollution-derived aerosol particles. This visible reminder of air pollution on hot days is one of the causes for the concern that air pollution will become much worse if temperatures rise as has been predicted in global warming scenarios.

Aerosol particles are not the only pollutant that occurs in hazy air. Dirty air tends to have high levels of many different pollutants and haze is often a visible surrogate for otherwise invisible pollutants which we can sense by smell, taste, or their effect on our eyes or breathing. A major concern is with ozone (O_3)

and other ingredients of photochemical smog which may likewise increase in concentration on a warmer planet. Our concerns are magnified by the robust statistical link between O_3 and temperature as revealed in monitoring data. A portion of this relation does appear to be cause and effect, whereby higher temperatures accelerate the rates of the chemical reactions generating photochemical smog. This, however, is not the whole story as both temperature and pollutant levels respond in common to changes in meteorological conditions such as increased sunshine. Distinguishing between the direct effects of temperature and the coincidental effects is critical to predicting the future.

This section will discuss the possible effects of a warmer climate on the air pollution problem in New York City focusing on ozone. We choose this focus for several reasons: Of the air pollutants regulated by the Environmental Protection Agency (EPA) (O_3, sulfur dioxide, nitrogen dioxide, carbon monoxide, lead, and particulate matter), the control of O_3 is the most problematic as judged from the limited success achieved from three decades of effort to lower O_3 levels. Significant adverse health effects have been attributed to O_3 at concentration levels common in urban atmospheres. Ozone is a component of photochemical smog, a combination of air pollutants that are formed in the atmosphere from chemical reactions involving nitrogen oxides (NO_x) and volatile organic compounds (VOCs, the principal components being various hydrocarbons). Most attention is directed towards ozone because of its impacts on respiratory function. However, many other ingredients of smog are eye irritants or noxious in other ways. Our understanding of the chemical and meteorological factors affecting the formation of O_3 and our past experience suggest that meeting the current federal standards will be very difficult, possibly requiring extreme emission control measures. Global warming is expected to exacerbate this situation.

In the following sections we describe the current O_3 pollution problem, the chemical and meteorological factors that are controlling O_3 levels, and ways in which these controlling factors might change due to global warming. We use a combination of monitoring data and model calculations to estimate the direct effect of a greenhouse-caused temperature perturbation on O_3 levels. Other climate variables will also change and have their impacts on O_3 levels, but we regard these non-temperature effects as being beyond our forecast capabilities. The sensitivity of O_3 to temperature, so derived, is used in a following section to estimate increased mortality due to increased O_3 from greenhouse warming.

Aerosol particles are also of concern because they are major components of acid rain, they are responsible for visibility degradation, and they have adverse health impacts. As with O_3, there is a statistical link between particle concentration and temperature. Although it is plausible that higher temperatures accelerate the rates of atmospheric reactions leading to aerosol products, this effect has not been quantified. Current evidence suggests that day to day variations in particle concentration are driven mainly by ventilation factors such as wind speed and wind direction relative to the location of emission sources. How these factors will change in a warmer climate cannot be predicted confidently. Accordingly we will not attempt to estimate the perturbation to aerosols occurring in a warmer climate. Instead, for comparison purposes, an estimate for the change in aerosol health effects will be derived based on the current relation between aerosols and temperature, realizing that this relation is likely a surrogate for the effects of other meteorological variables on particle concentration.

The Ozone Problem

The Federal Clean Air Act amendments of 1970, 1977, and 1990 classify O_3 as a criteria pollutant and specify the control of O_3 based on its acute respiratory effects to a susceptible population that includes the elderly, the very young, and asthmatics. The current federal standard requires that the fourth highest hourly average O_3 concentration recorded in a three-year period shall not exceed 120 ppb. Ozone also has deleterious effects at lower concentration including reductions in crop yields and damage to forests. Additional standards to protect against long-term exposure to moderate O_3 levels have been discussed but not implemented.

The EPA and corresponding agencies at the state level have been monitoring O_3 levels for more than three decades. In recent years (1982–1989) an average of 84 metropolitan areas, containing approximately half the population of the United States, reported O_3 levels in excess of the federal standard. Hourly concentrations close to 500 ppb have been reported in Los Angeles (1980) and 300 ppb in the New York area (1987).[3]

With three decades of progressively more stringent emission controls there has been a modest improvement in the urban O_3 problem, but O_3 levels in most large cities still violate the 120 ppb standard. Part of the reason that O_3 has not been susceptible to control measures is that over the same three decades there have been significant increases in population, industrial activity, and, especially, automobile traffic. Trends are difficult to determine because of year to year variability in meteorological conditions and the associated variability in O_3. Between 1980 and 1989, the maximum 1-hour O_3 concentration in urban areas decreased by 1.4% per year.[4] Ozone levels in 1988 were very high because of a particularly hot and dry summer over much of the United States; a trend based on the decade previous actually indicates an increase in O_3.[5] Violations of the O_3 standard in New York State for the period 1973 to 1992 are indicated in FIGURE 1. Most of the standard violations occur in the New York metropolitan area. Both the number of violations and their severity have been decreasing.

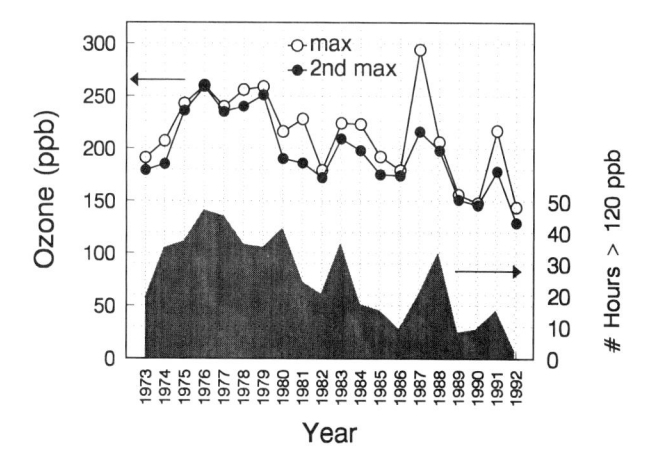

FIGURE 1. Ozone trends in New York State. (Source, NYSDEC, 1992.[3])

In recent years concern about high O_3 levels has been extended to non-urban areas even though these areas usually do not violate the 120 ppb standard. Based on a combination of routine monitoring, intensive field experiments, and model calculations, our current picture of photochemical smog is that it is often a regional scale phenomena. Under optimum conditions for O_3 production, concentrations higher than 80 ppb can occur over much of the eastern United States. Embedded within a regional event are hot spots corresponding to urban areas containing large NO_x and hydrocarbon emission sources. The Northeast Corridor, encompassing the cities of Washington, DC, Baltimore, Philadelphia, New York, and Boston, can have O_3 levels approaching and above the federal standard over its whole length. Ozone levels are usually at their highest 50 or more km downwind of major source regions. Thus, the highest O_3 recorded during 1994 in the northeastern United States was in Stratford, Connecticut (180 ppb). Ozone has a multi-day lifetime in the atmosphere and can thereby be transported significant distances. The plume from the Northeast Corridor is responsible for O_3 concentrations greater than 175 ppb observed at Acadia National Park, Maine. Further to the east, the plume has been observed over Yarmouth, Nova Scotia, causing an O_3 concentration in excess of 130 ppb.[6] With further eastward transport, the North American pollutant plume becomes part of the Northern hemisphere background. Man-made pollutants are thought to be responsible for more than half of the O_3 in the Northern hemisphere, even in remote regions.[7,8]

The regional O_3 background in the eastern United States is due to a combination of the melding together of plumes from multiple urban areas as well as O_3 generation in areas relatively remote from large emission sources. Because the air entering an urban area may contain an O_3 concentration that is a significant fraction of the federal standard, it becomes difficult (some would say impossible) for a city to control its own air pollution problem. The regulatory agencies are becoming more cognizant of this concern. Thus, the most recent amendment to the Clean Air Act mandates nation-wide controls for NO_x and VOCs from motor vehicles and NO_x from power plants. The 1990 amendment also establishes an interstate ozone transport region extending from Washington, DC to Maine, explicitly recognizing that standard violations can be due to upwind source regions.

Ozone Photochemistry

The essential ingredients to form O_3 in the atmosphere are NO_x, VOCs and sunlight. Note that we are restricting our attention to the troposphere; a different set of starting materials and reactions assumes primary importance in the stratosphere. NO_x and VOCs have both natural and man-made sources. In urban areas the man-made sources far outweigh the natural ones. This is often not the case in rural areas where emissions of isoprene and other very reactive hydrocarbons from trees can outweigh the man-made sources. Principal sources of hydrocarbons are motor vehicle exhaust, gasoline evaporation, and solvent evaporation. NO_x is emitted from high temperature combustion in motor vehicles and power plants.

Photochemistry is initiated by the absorption of a near UV photon splitting a chemical bond and forming a reactive free radical. Ozone itself is often the most important initiator producing a hydroxyl radical, (OH):

$$O_3 + h\nu \rightarrow O_2 + O(^1D) \qquad \textbf{(R1)}$$

$$O(^1D) + H_2O \rightarrow 2\ OH \qquad \textbf{(R2)}$$

Next follows a series of chain propagation steps which can involve reaction of OH with any of the thousands of VOCs present in the ambient atmosphere. A simple example involves reaction of CO.

$$OH + CO \xrightarrow{O_2} HO_2 + CO_2 \quad \text{(R3)}$$

$$HO_2 + NO \rightarrow NO_2 + OH \quad \text{(R4)}$$

$$NO_2 + h\nu \rightarrow NO + O \quad \text{(R5)}$$

$$O + O_2 \rightarrow O_3 \quad \text{(R6)}$$

The net effect of this sequence is that CO is oxidized to CO_2 and that one molecule of O_3 is created. VOCs (using CO as an example) are in essence the fuel that runs the photochemistry. NO_x is not removed in the above cycle and can be regarded as a catalyst. Chain termination occurs by reactions of free radicals with each other, for example,

$$2\ HO_2 \rightarrow H_2O_2 + O_2 \quad \text{(R7)}$$

and more importantly in urban areas by reactions of NO_x with free radicals such as

$$OH + NO_2 \rightarrow HNO_3. \quad \text{(R8)}$$

The last reaction removes NO_x from the active photochemistry and thereby necessitates a continued supply of NO_x for the photochemistry to proceed. NO_x and free radicals can be temporarily removed from the active photochemistry via the formation of peroxyacetyl nitrate (PAN).

$$CH_3CO_3 + NO_2 \rightarrow PAN \quad \text{(R9)}$$

The above reactions are a schematic outline of the 100 or so reactions typically treated in computer models of the polluted lower (altitude) atmosphere. Such models also treat the transport of pollutants, necessitating a description of wind fields and a geographically accurate map of emission sources. Depending on application, the spatial modeling domain can cover an urban area, the entire planet (albeit with less detail in the chemistry) or something in between. Much of what we know about the generation of photochemical smog is the result of computer calculations coupled with observations of the ambient atmosphere or coupled with controlled smog chamber experiments.

Control strategies are based in large part on model calculations. Typically a model is run for a high O_3 episode for which there are observations to compare against. Having demonstrated skill in predicting the observed O_3 levels (and less often concentrations of other photochemically active substances), the model is then rerun with different levels of NO_x and VOC emissions corresponding to plausible (and in some cases, implausible) control strategies. Results are often displayed as isopleth maps showing the maximum O_3 produced in an air parcel during the course of a day as a function of NO_x and VOC emission rates (or sometimes as a function of early morning concentrations of NO_x and VOCs). A

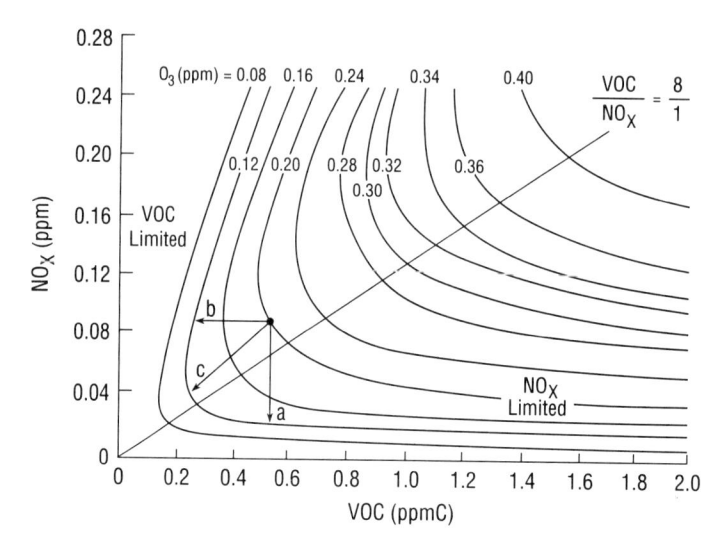

FIGURE 2. An example of an O_3 isopleth map, showing predicted O_3 concentration as a function of NO_x and VOC. (Adapted from Dodge.[33])

schematic example is given in FIGURE 2. Arrows on the diagram indicate three possible control strategies that would result in meeting a 120 ppb standard starting from a point with an O_3 concentration of 200 ppb. In this case the standard could be met with control of NO_x, control of VOC, or combined control of NO_x and VOC.

Ozone isopleth graphs, while specific to individual cities, all have common generic features. Ozone production occurs most efficiently when the ratio of VOC to NO_x is about 8 to 10. At a very high ratio, the atmosphere is NO_x limited. Ozone production is then insensitive to the amount of VOC; only NO_x controls are effective. At a very low ratio, the atmosphere is VOC limited. Ozone production can then be controlled by reducing VOC, but reductions in NO_x can in some extreme cases lead to more O_3. NO_x limited and VOC limited regions are shown in FIGURE 2.

The relative amounts of NO_x and VOCs emitted and in the air is thereby an important piece of information in deciding on an O_3 control strategy. Regulatory efforts have focused largely on control of VOC emissions. There has been vigorous debate as to whether this is an optimum policy.[9] Recent evidence shows that most of the United States is in a NO_x-limited condition most of the time. VOC-limited conditions are still expected in major urban areas, especially at times leading to peak O_3 levels. Sillman[7] notes the difficulty that this situation poses for the regulatory agencies. The optimum strategy for an individual city is often to lower O_3 production in the city by reducing VOC emissions while persuading the neighboring upwind regions to reduce NO_x emissions in order to lower O_3 in the "background" air entering the city.

Meteorological Factors

FIGURE 2 presents information on the response of O_3 to changes in its chemical precursors, NO_x and VOCs, but leaves unsaid the fact that O_3 concentrations depend on a host of other variables. Primary factors are ventilation, sunlight, temperature, humidity, rainfall, and wind direction vis-á-vis the location of upwind emission sources. In areas with poor ventilation (low wind speed and low mixing height) there is a continued input of pollutants to an air mass, resulting in high concentrations of emitted substances as well as high concentrations of their reaction products. Stagnant conditions in the summer usually co-occur with high humidity, temperature, and solar intensity and tend to produce high concentrations of O_3 and other reaction products. In many instances wind direction is a controlling variable. With winds from one direction, an ocean could be upwind, while from another direction, an urban area could be upwind.

Effect of Gobal Climate Change on O_3

Rising CO_2 levels are expected to perturb the climate of the earth. For the purpose of this planning exercise we rely on the scenarios generated for the *Metropolitan New York in the Greenhouse* workshop (A. J. Broccoli, personal communication, based on the IPCC report[10]) which specify a best estimate global annual average temperature rise of 1.1°C in the year 2030 and 2.4°C in the year 2070. The wintertime and summertime temperature rise in New York are predicted to be 1.5 to 3°C and 1 to 2°C, respectively. By 2070, the temperatures are predicted to rise by 3 to 6°C in the winter and 2 to 4°C in the summer.

There are multiple ways in which a temperature change of this magnitude coupled with changes in other climate variables could have an impact on atmospheric chemistry in general and formation of O_3 in particular. Mechanisms suggested include:

1. An intrinsic dependence of rates of atmospheric chemical reactions on temperature. Oxidation reactions of OH with hydrocarbons (schematically indicated by **R3**) proceed faster at high temperature. PAN formation **(R9)**, which inhibits the production of O_3 by removing NO_x and radicals, becomes less significant at high temperature.
2. Changes in emission rates in response to increases in temperature. Anthropogenic emissions of NO_x and hydrocarbons would increase (all other things being equal) due to increases in solvent evaporation, motor vehicle use, and electric generation. Natural emissions of hydrocarbons from trees are a steeply increasing function of temperature. The large contribution of natural hydrocarbons to O_3 generation over much of the eastern United States is one of the major reasons that O_3 control policies, based on restricting emissions of anthropogenic VOCs, have been less successful than anticipated.
3. Changes in atmospheric humidity. Water vapor is a reactant in the photochemical smog system leading to the production of reactive free radicals **(R2)**.
4. Changes in cloud cover, which affect solar intensity, which in turn affects the rates of photolysis reactions (*i.e.*, **R1** and **R5**).
5. Changes in precipitation patterns. The air is almost always cleaner during and after rain. This results from the removal of some pollutants, most notably aerosol particles which are responsible for visibility degradation. Precipitation is also associated with good ventilation and reduced sunlight.
6. Changes in wind patterns. Of particular importance are the frequency and severity of stagnation conditions that allow for the multi-day buildup of

air pollutants. Changes in the patterns of prevailing winds could bring air that is more or less polluted to a particular receptor site.
7. Changes in the height of the mixed layer (the layer of the atmosphere above the surface of the earth, typically extending to an altitude of about 1 km in which there is relatively vigorous mixing) making available a greater or lesser volume for dilution.

The global climate models that are used for predicting the temperature effects of increased greenhouse gases also yield predictions for the changes in other climatic variables such as precipitation, storm tracks, and cloud cover. However, we regard these as too uncertain to base a planning exercise on.

We can quantify the effects of a warmer climate on O_3 levels in two ways: by examining the historical relation between O_3 and temperature and by performing model calculations in which temperature or other climatic factors are varied. The historical relation, based on coincident observations of temperature and O_3, is the more direct approach but does not tell us whether temperature itself is the controlling factor or whether both temperature and O_3 are varying in response to another factor such as sunlight intensity. This is a critical question since we are hypothesizing a future in which temperature increases but, for example, solar intensity may remain constant or even decrease. Models have their own uncertainties but have the advantage of being able to address the effects of several factors independently.

Observations

Monitoring data indicate that O_3 concentration tends to increase with temperature, the effect being most pronounced at temperatures above 27°C. FIGURES 3 and 4 illustrate the relation between daily maximum values of O_3 and temperature for New York City and for a set of four nonurban locations. There is considerable scatter to the relation between O_3 and temperature, reflecting

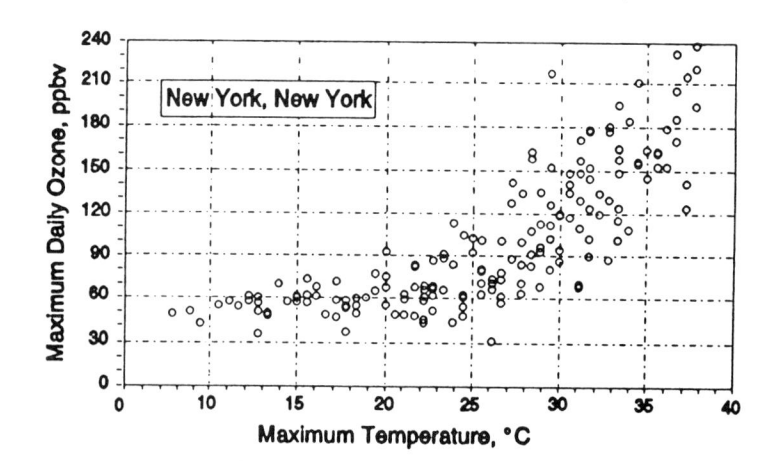

FIGURE 3. Daily maximum O_3 concentration as a function of daily maximum temperature in New York City during the months from May to October, 1988 to 1990. (Source, United States EPA.[11])

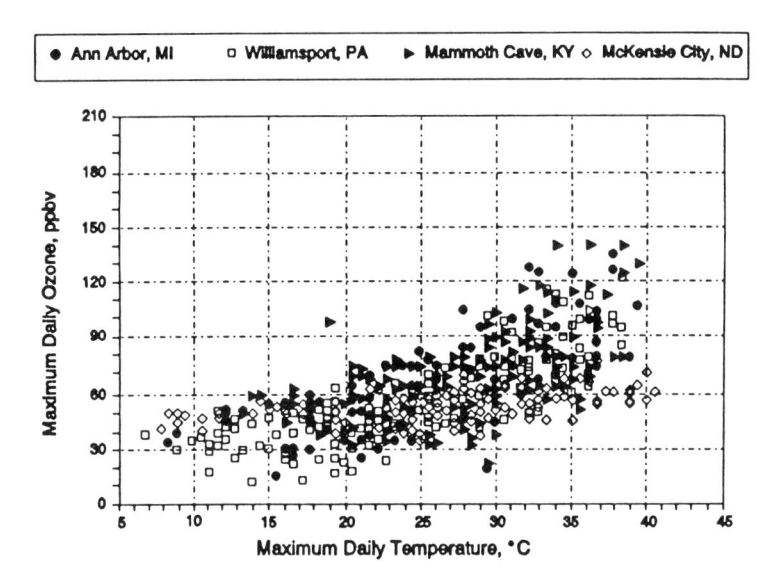

FIGURE 4. Daily maximum O_3 concentration as a function of daily maximum temperature at four nonurban sites during the months from May to October, 1988 to 1990. (Source, United States EPA.[11])

the influence of other chemical and meteorological factors such as emission rates, sunlight, and wind speed. As noted in the EPA Criteria Document,[11] there is a well-defined upper bound on O_3 concentration that reflects the maximum concentration achieved under optimum conditions. This maximum is seen to increase with temperature. Statistics for several locations are summarized in TABLE 1. Jones et al.[12] present another view of the relation between temperature and O_3 by comparing the number of violations of the federal O_3 standard with the number of days in which the temperature was above 90°F, for several cities including New York City. As noted in the National Research Council report,[9] there is a strong correlation between these two variables. There are also strong correlations between temperature and other air pollutants as indicated in FIGURE 5 for aerosol particles in Philadelphia.

A straightforward explanation of the data in FIGURES 3 and 4 is that maximum O_3 concentrations do indeed depend in a causal way on temperature. However, it is likely that other factors contribute to this relation, in particular for regional high-O_3 episodes over the eastern United States. Jacob et al.[13] note that regional episodes appear to be driven primarily by stagnation conditions. Low wind speeds allow for the continued addition of emitted NO_x and hydrocarbons to an air mass and cause O_3 to build up over a multi-day period. Jacob et al. further note that the correlation between O_3 and temperature may reflect the dependence of temperature on air mass origin or solar radiation. Part of the association between temperature and O_3 in New York City may be due to the circumstance that warm air generally arrives from the south or southwest, from regions that have high emission rates of O_3 precursors and high O_3 concentrations. This association is also thought to explain much of the dependence of aerosol concentration on temperature.

TABLE 1. Observed Dependence of Peak O_3 on Diurnal Maximum Temperature (ppb°C^{-1}) between April 1 and September 30, 1988

Location	$\Delta O_3/\Delta T$ T < 27°C	T > 27°C
Urban sites:		
NY-NJ-CT	1.5	8.8
Detroit	1.4	4.4
Atlanta	3.2	7.1
Phoenix	—[a]	1.4
Southern Calif.	11.3	—[a]
Nonurban sites:		
Williamsport, PA	1.2	4.0
Saline, MI	0.8	3.1
Mammoth Cave, KY	—[b]	4.4
Kentucky, cleanest site	—[b]	3.4
Williston, ND	—[b]	0.8
Billings, MT	—[b]	0.7
Medford, OR	0.5	3.3

[a] Too few cases to be included.
[b] Trend not statistically significant.
SOURCE: Sillman & Samson, 1994.[16]

Models

Morris et al.[14] applied a regional photochemical model to the study of O_3 episodes in central California and the Midwest/Southeastern United States. Results of increasing temperature by 4°C with an attendant increase in water vapor varied from day to day and place to place. In California the daily maximum O_3 concentration increased by 2 to 20%, while in the Midwest/Southeast the change varied from −2% to 8%. In both studies there were large increases in the population exposed to O_3 levels above the 120 ppb standard. The sensitivity of O_3 to temperature in New York City was tested by Gery et al.[15] using an "EKMA" model

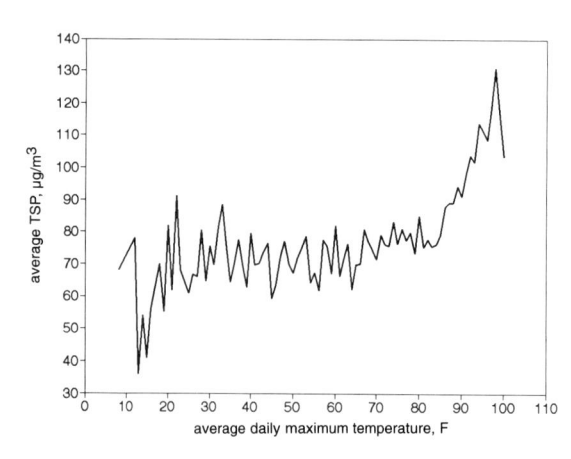

FIGURE 5. Relationship between particle concentration and daily maximum temperature in Philadelphia, 1973–1980. (Adapted from Wyzga and Lipfert, 1994.[26])

similar to that used to generate the isopleth map shown in FIGURE 2. A 2°C temperature rise was found to cause an increase in O_3 from 125 to 130 ppb. Similar results were found for Philadelphia and Washington, DC. A modeling study by Sillman and Samson[16] predicts that approximately half of the observed increase of O_3 with temperature in Detroit and a rural site in Michigan can be accounted for by a combination of temperature-dependent emissions of natural hydrocarbons and temperature-dependent rate constants. A mechanistic analysis indicates that the temperature dependence results primarily from the formation of PAN tying up free radicals and NO_x as the temperature decreases, thereby inhibiting O_3 production under NO_x limited (rural) and VOC limited (urban) conditions.

Although the three simulations cited above are for different regions and incorporate somewhat different assumptions on what else changes when temperature is increased, they agree to the extent that they predict urban O_3 increases of about 3 ppb per degree C temperature increase. This is about 1/3 of the observed O_3 trend in NYC, ancillary changes in meteorological conditions accounting for the other 2/3. Our sensitivity estimate is based on starting with about 120 ppb of O_3. Additional calculations are needed to determine if a larger increase in O_3 occurs if the base case concentration is greater. We expect a temperature rise to change the rates of hydrocarbon reactions and the availability of NO_x, effectively mimicking emission increases of these substances. This will shift our position on the O_3 isopleth map, changing the optimum emission control strategy. The magnitude and even direction of such a shift cannot be determined without detailed calculations specific to the New York metropolitan area.

A postulated 2°C increase in temperature, according to the above analysis, translates into a 6 ppb increase in O_3. Given the year to year variations in O_3, such as displayed in FIGURE 1, it is not likely that a change of this magnitude would be evident from monitoring data. Other changes in climate variables could make this figure lower or higher.

THE HEALTH EFFECTS OF CLIMATE CHANGE

It has long been known that extreme weather can have extreme effects on health, and through the centuries man has developed effective means of protection. Further, populations have been shown to adapt to their climates in various ways, including design of shelters and clothing and through personal habits. Thus, it is to be expected and analysis has shown that perturbations about the normal weather patterns carry larger risks than do steady gradual changes in climate. For example, a heat wave occurred in Athens in July 1987, with daily mean temperatures around 35°C (95°F) for six days with a peak daily maximum value of 41°C (106°F).[17] This was an increase of only about 7–9°C over normal levels, but the daily death rate increased by an order of magnitude, from an average of about 23 to over 200 deaths per day. Yet, such temperature levels are experienced routinely in many parts of the world, including the United States. In New York in 1993, for example, a peak temperature of 102°F was reached with a mean of 91°F, over a 3-day period. If daily deaths increased during this period in New York, the increase went largely unnoticed.

The first section of this paper indicated that changes in climate may also bring changes in ambient air quality, both directly and indirectly. Direct changes may result from changes in atmospheric reaction rates, for example. Indirect effects may result from changes in atmospheric transport and dispersion factors, and from increased pollutant emissions resulting from man's attempts to escape the heat.

Health Effects of Air Pollution and Weather

The early analyses of the most severe health effects of air pollution, *i.e.*, premature mortality and respiratory hospitalizations, tended to focus on winter episodes.[18] In winter, air quality problems tend to be caused by primary emissions from space heating and by reduced atmospheric dispersion during stagnation episodes. Such winter events may coincide with the periods of peak mortality and the presence of infectious respiratory diseases, such as influenza and peneumonia. However, since the Clean Air Act of 1970 and its progeny, space heating fuel quality has improved markedly, especially in New York City, to the point where summer may be the period of peak air pollution. In summer, secondary pollutants including ozone and fine particles are the focus of concern. This not only changes the chemical focus, but also the locations of concern, since the ambient concentrations of secondary pollutants usually peak well downwind of their sources, say *ca.* 50–100 km or more, and the pollution becomes dispersed over a wide area. Also, in hot weather, many people spend relatively more time outdoors, thus increasing their exposure to air pollution. This may be especially true of those lacking access to air conditioning.

The numbers of deaths officially classified as "heat related" may substantially underestimate the true effects, in part because of the uncertainties of cause-of-death coding. For example, in Philadelphia, not only is a hot environment required but the decedent must also be elderly or infirm to be listed under this cause of death.[19] During 1979–1988, only 4523 deaths or about 0.023% of all deaths in the United States were officially attributed to "excessive heat exposure," and 1700 of those were in one year. Statistical analyses of the timing of death (described below) tend to find much larger effects of increased ambient temperatures. Interestingly, air conditioning is recommended as a protective measure, but not fans.[19] Increased ventilation of air above skin temperature would exacerbate heat stress, and bringing outside air into the home could add to air pollutant exposure. One hundred and eighteen deaths were attributed to heat stress over a 9-day period in Philadelphia (about 26% excess);[20] this heat wave was also experienced in New York City, where the peak temperature reached 102°F and ozone levels also peaked.

The health effects of these excursions in environmental conditions are typically analyzed using the methods of time-series analysis.[21–30] The basic technique involves various types of multiple regression analysis, in which daily counts of deaths, hospital admissions, or other indices of health are the dependent variable. Independent variables include daily measures of air quality and other temporal factors which might be correlated with health and air quality. These include weather variables, of which temperature is usually the most important, and other perturbations such as days of the week, holidays, and influenza epidemics. It is also important to account for long-term trends that might confound the analysis. These include both secular trends and seasonal cycles. In the former, it is possible that coincident trends in improved health, resulting from better medical care or healthier lifestyles, for example, might be (incorrectly) associated with a concurrent trend in improved air quality resulting from imposition of emissions controls. Seasonal cycles may confound, for example, if the expected winter peak in mortality or respiratory illness (resulting from infectious disease cycles) is inappropriately associated with the seasonal cycle in air quality that results from increased space heating emissions and reduced atmospheric dispersion in winter.

Epidemiologists have developed two major techniques for distinguishing the short-term effects of air pollution from those of these potential confounders.

Schwartz and his colleagues,[25] for example, use a number of auxiliary independent variables as controls. These include trend variables and dummy variables for extreme events like heat waves; daily temperature is used primarily as a control for seasonal cycles. In general, this method does not yield a regression coefficient for hot weather effects as a function of maximum temperature. Other investigators[22,26,27] have used the classical time-series methodology called "filtering," in which new "perturbation" variables are created by subtracting the appropriate running mean from each original variable. Daily temperature is used as a predictor of daily health status in this technique, rather than as a seasonal control. Filtering absolutely removes the long-term trends, but the regression coefficients may be sensitive to the details of the filtering process.[24] Comparison of these two basic regression techniques suggests that filtering tends to assign less of the observed health variance to air pollution, which has been interpreted as a failure of the first method to control sufficiently for long-term trends and seasonal cycles.[26]

It is important with all analyses of daily variations in health status to account for lags between exposure and response; plots of response versus lag may help us interpret the regression results. For example, the response must lag the exposure if the association is to be regarded as causal. In addition, the degree of prematurity of response may be assessed by considering whether initial positive responses may be effectively compensated by subsequent negative responses, such that the sum over a longer time period approaches zero. In mortality analysis, this compensation process has been referred to as "harvesting" or mortality "displacement."[2] FIGURES 6–8 compare some of these lag plots. In FIGURE 6, hospital admissions in Southern Ontario in July–August[23] increase with ozone and sulfate aerosol beginning on the day after exposure and show no indication of a subsequent decrease (perhaps longer lags should have been examined). In FIGURES 7 and 8, daily mortality on hot days in Philadelphia also increases with ozone or particulates on the day after exposure and then decreases on subsequent days. In contrast, the effects of temperature are felt immediately (FIGURE 9), and show subsequent decreases, more so for those under age 65.

These air pollution effects have been shown to exist at air quality levels well within current state and federal standards; thus exceedances of those standards is not an appropriate criterion for evaluating the severity of health effects. The federal government is currently considering lowering these standards, but it seems unlikely that this type of health risk can ever be completely eliminated. The question that naturally arises is the extent to which air conditioning might help: air conditioning not only protects against atmospheric heating, but, given appropriate air filters, it can also protect against outside air pollution. Rogot et al.[31] estimated that mortality during "hot" weather (average temperature > 70°F) was 42% lower nationwide for those having central air conditioning (A/C), compared to those with no A/C, based on a nationwide sample of 2275 deaths. Their analysis used mortality during "non-hot" weather as controls to preclude socioeconomic confounding. Rogot et al.[31] found A/C to be of greatest benefit in Florida. Without Florida, the national mortality benefit dropped to 29%; their analysis showed no A/C benefit in New York, but there were only 11 New York deaths during hot weather in this sample. However, Kalkstein[2] cited an estimate of 21% savings in heat-related deaths in New York due to A/C, using a different analysis technique.

In the analysis of Rogot et al.,[31] the benefit attributed to A/C could include relief from both heat and from outside air pollution and we regard 21% as an upper limit estimate for New York. The Philadelphia statistical analysis[26] probably provides a better basis for estimating effects in New York, owing to the proximity

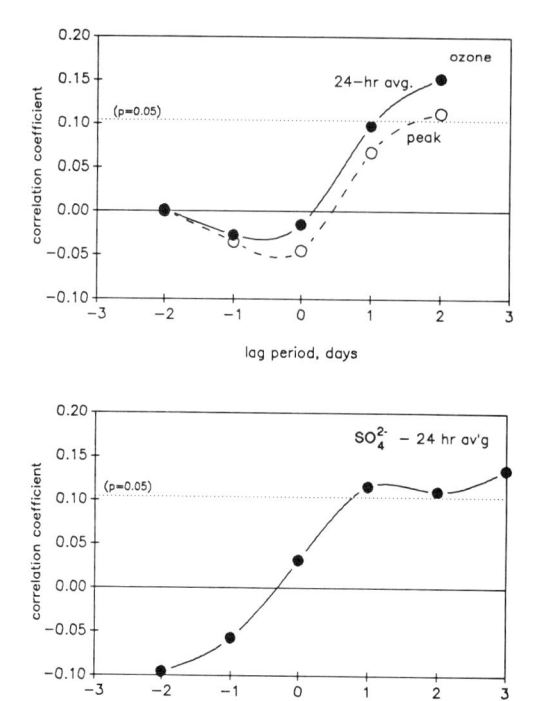

FIGURE 6. Correlations between hospital admissions for respiratory causes and air quality in Southern Ontario; (a) ozone; (b) sulfate aerosol. (Adapted from Lipfert and Hammerstrom.[23])

of the two cities, although Philadelphia's summers tend to be slightly hotter than New York's.

Estimated Effects of Global Warming in New York City

In the first section of this paper, an increase of 6 ppb ozone over a baseline of 120 ppb was predicted, based on a temperature increase of 2°C. We will use

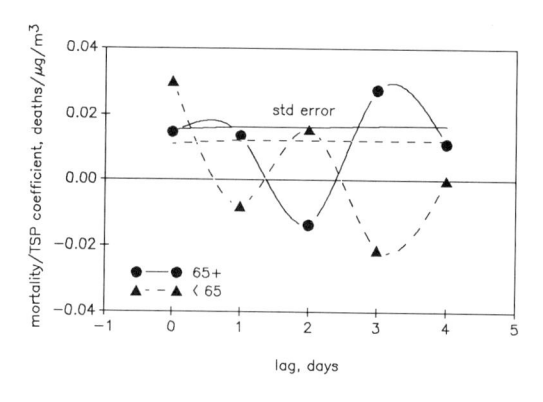

FIGURE 7. Regression coefficients for ozone on the deviations of daily death counts (● = persons over 65; ▲ = persons under 65) from a 15-day moving average, with various covariates included, on days with maximum temperature of 85°F or more. (Adapted from Wyzga and Lipfert.[26])

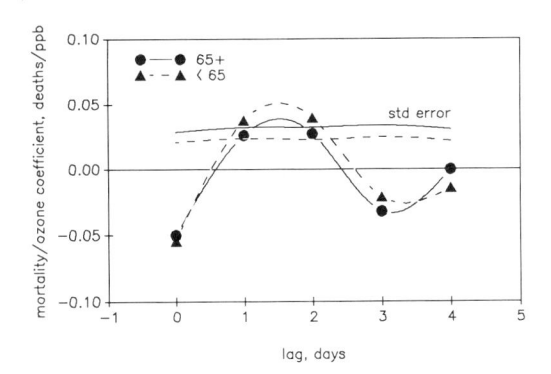

FIGURE 8. Regression coefficients for TSP on the deviations of daily death counts from a 15-day moving average, with various O_3 and temperature, on days with maximum temperature of 85°F or more. (Adapted from Wyzga and Lipfert.[26])

these estimates to predict changes in daily mortality in New York City, assuming no changes in pollution emissions. Although air pollution and especially ozone can have many different types of health effects, we use premature mortality as an index, in part because of its irreversible nature and in part because of the availability of the required coefficients for the effects of heat and of air pollution.

Mendell et al.[32] found an average summer mortality/temperature coefficient in several different locations of 0.45% per degree F (0.81%/°C); however, these studies may have neglected the harvesting effect mentioned above. In addition, graphical analysis shows that the effects of temperature in many cities tend to increase dramatically after some threshold is crossed, typically around 33°C (91°F). Thus, hot days should be examined separately to gain a better understanding of the health responses. Wyzga and Lipfert[26] looked at days with maximum temperatures over 85°F in Philadelphia, as a subset of all days from 1973–1980. Their analysis combined lags up to 4 days in order to allow for mortality harvesting. On a year round basis, for example, using the Philadelphia analysis for all days, a 2°C temperature increase would increase daily mortality by 0.45%, or by about 340 deaths per year in New York City. Computed using the regression results for "hot" days alone (assumed to be 18% of the total), the effect of 2°C would be about 3.4%, or about 470 deaths per year. This exercise suggests that mortality changes due to temperature may be neglected in other seasons.

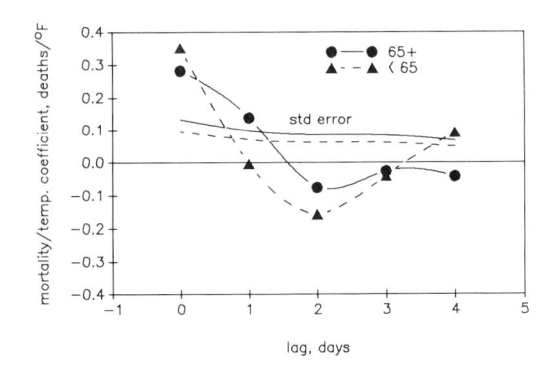

FIGURE 9. Regression coefficients for daily maximum temperature on the deviations of daily death counts from a 15-day moving average, with TSP and O_3 included, on days with maximum temperature of 85°F or more. (Adapted from Wyzga and Lipfert.[26])

Interactions between summer ambient temperatures and air pollution have been shown in several cities, including New York (see FIGS. 3–5). After filtering to remove trends, Thurston et al.[28] showed positive correlations in Toronto from 0.38 to 0.68 between maximum daily temperature and ozone, suspended particulate matter, sulfate aerosol, NO_2, SO_2, and aerosol acidity. Lipfert and Hammerstrom[23] showed similar correlations there based on the unfiltered data. Schwartz[29] showed temperature correlations of 0.67 and 0.32 for ozone and particulate matter of 10 microns or less (PM10), respectively, in Detroit. In their study of summer hospital admissions in New York State, Thurston et al.[30] found positive bivariate correlations between daily maximum temperature and hospital admissions, as well as with ozone, sulfate aerosol, and aerosol acidity.

In order to estimate the effects of the air pollution increases that would accompany increases in ambient temperature, regressions are required that estimate the temperature and pollution effects jointly (along with any other potentially confounding variables). The ozone-mortality coefficients from the analysis of the filtered Philadelphia data[26] were 0.008 for those under 65 and 0.002 for those 65 and over on a year-round basis, expressed as percent change in mortality for a 1% change in ozone. An increment of 6 ppb over a baseline of 120 ppb is an increment of 5%, which then translates into a mortality change of about 0.02% or about 16 deaths per year. The mortality effects of ozone were not stronger on hot days, perhaps because of the competing direct effect of temperature.

However, of the hospitalization studies, only Burnett et al.[27] provided a regression coefficient for temperature; their results indicated that the 2°C temperature increase from global warming would increase respiratory hospital admissions by about 0.5% and the accompanying ozone (6 ppb) would create an additional 0.6% increase in admissions, for the summer period.

The health effects of increased airborne particles should also be taken into account in these estimates, which requires separation of the effects of daily perturbations from seasonal and long-term effects. The association between mortality and particulates was shown by Wyzga and Lipfert[26] to increase dramatically on hot days; the relationship between total suspended particulates (TSP) and temperature is also much stronger on hot days in Phildelphia. On a year-round basis, an increase of 2°C would correspond to a TSP increase of 0.5 $\mu g/m^3$ or about 0.7%, based on a regression model which also accounted for wind speed, precipitation, relative humidity and change in barometric pressure. This would result in an increase in mortality of 0.01% for those aged 65 and over. For days with maximum temperatures of 85°F or more, which amounted to about 18% of the total, an increase of 2°C would correspond to an increase of 2.8 $\mu g/m^3$ or about 3% on those days. This would result in an increase in mortality of 0.05% for those under 65 and 0.32% for those 65 and over. These figures result in a weighted year-round increase in total mortality of about 0.04%. The fact that hot days account for a disproportionate share of the total effect implies a nonlinear relationship, which was confirmed by investigating other ranges of daily maximum temperature. TSP was not significantly associated with temperature in any other (lower) temperature range, and the slope became negative below about 40°F, apparently because lower temperatures in winter imply higher emissions from space heating. The results of these calculations are summarized in TABLE 2. (Note that we have disregarded the lack of statistical significance of the effects of air pollution on daily mortality seen in FIGURES 7 and 8, for the purpose of making these estimates.) Based on the 1987 death count for New York City of about 76,250, the estimated total effect of global warming would result in an annual increase of about 511 deaths, 90% of which are due to the effects of temperature. The combined estimate for respiratory

TABLE 2. Health Effects of a Temperature Increase of 2°C

	Mortality	Respiratory Hospital Admissions
From temperature alone:	0.61%[26]	0.45%[27]
From ozone:	0.02%[26]	0.58%[27]
From airborne particles:	0.04%[26]	0.32%[29]
Total effect:	0.67%	1.3%

hospital admissions[30] would be about 200, based on summer days only. Premature mortality thus appears to be the more important health impact, in terms of both severity and frequency.

Mitigation Options

Air Conditioning

In theory, it should be possible to mitigate a large portion of the effects of global warming through air conditioning. We assume that most of the excess deaths occur to those who are already sick, and therefore we use residential air conditioning as the paradigm. These estimates are necessarily crude and are presented here only for the sake of illustration and discussion. In order to provide a worst-case illustration, we assume that the additional energy is to be supplied by new fossil-fueled power plants located in the metropolitan area.

Some handbooks estimate that about 0.003 ton of air conditioning or 36 Btu/h is required for each additional square foot of conditioned space. We assume that efficiencies have increased over time and take 2 tons of refrigeration per household or 1 ton per person as a rough estimate for New York City. This requires an electrical capacity of about 2 kw per person, and electrical system planning data indicate that about 2000 kwh per household-year is required in New York for central air conditioning (personal communication, P. Coffey, 1994). Taking the required additional market penetration of residential air conditioning at 50%,[a] we estimate that about 7500 MW of new electrical generating capacity would thus be required, or an approximate doubling of the present Con Edison system. At $0.15 per kwh, the additional seasonal operating cost would be around $300 per household or about $560 million, to which the costs of capital for air conditioning and generation should be added (increases in electric rates as a result of new plant construction are not included in this estimate).

The additional air pollution from this electrical generation must also be considered. We estimated emission increases of about 4130 tons of SO_2 and NO_x (each) and about 1030 tons of particulate matter over a 100-day period of air conditioning. If emitted from tall stacks, these emissions would amount to about 1.7 and 0.43 $\mu g/m^3$ at ground level, respectively, or increases of about 5% or less over present background concentrations. The increased NO_x emissions (about 5% on a summer day) would probably also contribute to the formation of additional O_3 downwind of the city. These increases are of the same order as the corresponding global warming increases (which of course would still be there).

The use of air conditioning can protect against all heat waves and most of the outside air pollution, not just the increases from global warming. This implies that

---

[a] See note added in proof on page 110.

many additional lives would be prolonged, beyond those associated with the global warming increase. Since economists often use several million dollars as the value of a "statistical life," a policy of increased residential air conditioning seems cost effective. This is likely to be the case even when other types of environmental costs are considered.

Trading Reduced Air Pollution against Increased Heat

Additional air conditioning will result in increased CO_2 emissions (about 3.5 million tons in the example above) if the energy is supplied by fossil fuels. This will provide a positive feedback effect which will tend to worsen global warming along an accelerating path, although the increase in CO_2 will be small because the additional energy is only needed during hot weather, not year round. It has been suggested that a better policy might be to reduce air pollution by imposing additional controls on emissions or fuel use, such that the increased health effects of heat are offset by better air quality. Such a scheme could have a negative feedback effect if overall fuel use is reduced as a result of the policy.

The magnitudes of the air pollution reductions required may be estimated from the figures given in TABLE 2. The leverage of air pollution effects with respect to temperature is less for mortality, which we use as the basis for this example. If we attempt to use ozone reductions to compensate for the effects of heat waves, the reduction would have to be about 30 times the increment used in TABLE 2 or about 150 ppb. Since the baseline level was 120 ppb, this scheme is clearly impractical. However, the compensating reduction that would be required from particulate matter is more modest, about 15 times the increment shown in TABLE 2 or 7.5 $\mu g/m^3$ on a year-round basis or 42 $\mu g/m^3$ on the hottest days (a reduction of about 30%). While such a tradeoff scheme may seem feasible in theory, one must realize that this is an attempt to trade "statistical lives" and that compensation may not occur on an individual basis. Heat and pollution events will not always coincide, and individuals may respond differently to different environmental insults. We have no guarantee that easing the overall particulate burden would in fact protect someone who happens to be especially sensitive to heat.

Concluding Discussion

This analysis has shown that the air quality and health effects to be expected from global warming in New York City appear to be relatively modest, based on ozone and particulates as indices of air quality and mortality as an index of health effects. However, there are other considerations. The analysis is based on a uniform temperature increase, while the actual mortality effects of heat seem to result from daily perturbations. Unfortunately, we have no predictions of changes in the variability of conditions under global warming. Also, the analysis does not account for downwind effects in suburban or rural areas, and there are other types of health effects to consider, such as respiratory symptoms, some of which are more responsive to ozone. Effects of a less severe nature tend to apply to a larger segment of the population and thus could be more important than the mortality and hospitalization paradigms used here.

With regard to mitigation options, it seems clear that increased use of air conditioning can prolong lives and that reductions in fuel use may not only prolong lives but also mitigate the increase in global warming (if implemented on a sufficiently large scale). The challenge for the future is to find ways to accomplish both of these goals.

REFERENCES

1. HAINES, A., P. R. EPSTEIN & A. J. MCMICHAEL. 1993. Global health watch: Monitoring impacts of environmental change. Lancet **342:** 1464–1469.
2. KALKSTEIN, L. S. 1993. Direct impacts in cities. Lancet **342:** 1397–1399.
3. NEW YORK STATE DEPARTMENT OF ENVIRONMENTAL CONSERVATION. 1992. New York State Air Quality Report Ambient Air Monitoring System. DEC Publication, Annual 1992 DAR-93-1.
4. EPA (ENVIRONMENTAL PROTECTION AGENCY). 1991. National Air Quality and Emissions Trend Report, 1989. EPA Report 450/4-91-003, Research Triangle Park, NC. February.
5. EPA (ENVIRONMENTAL PROTECTION AGENCY). 1990. National Air Quality and Emissions Trend Report, 1988. EPA Report 450/4-90-002, Research Triangle Park, NC. March.
6. KLEINMAN, L., P. DAUM, S. SPRINGSTON, J. LEE, Y.-N. LEE, X. ZHOU, R. LEAITCH, C. BANIC, G. ISAAC, I. MACPHERSON & H. NIKI. 1994. Trace gas measurements over southern Nova Scotia during the 1993 NARE. Presented at the Joint 8th CACGP Symposium/2nd IGAC Conference, Fuji-Yoshida, Japan, September 5–9, 1994.
7. SILLMAN, S. 1993. Tropospheric ozone: The debate over control strategies. Annu. Rev. Energy Environ. **18:** 31–56.
8. PARRISH, D. D., J. S. HOLLOWAY, M. TRAINER, P. C. MURPHY, G. L. FORBES & F. C. FEHSENFELD. 1993. Export of North-American ozone pollution to the North Atlantic Ocean. Science **259:** 1436–1439.
9. NATIONAL RESEARCH COUNCIL. 1991. Rethinking the Ozone Problem in Urban and Regional Air Pollution. Washington, DC: National Academy Press.
10. IPCC (INTERGOVERNMENTAL PANEL ON CLIMATE CHANGE). 1990. Climate Change: The IPCC Scientific Assessment. J. T. Houghton, G. J. Jenkins, and J. J. Ephraums, Eds. Cambridge: Cambridge University Press.
11. ENVIRONMENTAL PROTECTION AGENCY. 1993. Air Quality Criteria for Ozone and Related Photochemical Oxidants. Vol. 1. EPA Report 600/AP-93/004a; December, Research Triangle Park, NC.
12. JONES, K., L. MILITANA & J. MARTINI. 1989. Ozone Trend Analysis for Selected Urban Areas in the Continental U.S. Paper 89-3.6. Presented at the 82nd Annual Meeting and Exhibition of the Air and Waste Management Association, Anaheim, CA. June 25–30.
13. JACOB, D. J., J. A. LOGAN, G. M. GARDNER, R. M. YEVICH, C. M., SPIVAKOVSKY & S. C. WOFSY. 1993. Factors regulating ozone over the United States and its export to the global atmosphere. J. Geophys. Res. **98:** 14,817–14,826.
14. MORRIS, R. E., M. W. GERY, M. K. LIU, G. E. MOORE, C. DALY & S. M. GREENFIELD. 1990. Examination of the sensitivity of a regional oxidant model to climate variations. *In* The Potential Effects of Global Climate Change on the United States. J. B. Smith and D. A. Tirpak, Eds.: 510–517. New York: Hemisphere Publishing.
15. GERY, M. W., R. D. EDMOND & G. Z. WHITTEN. 1987. Tropospheric Ultraviolet Radiation: Assessment of Existing Data and Effect on Ozone Formation. EPA Report 600/3-87/047; October, Research Triangle Park, NC.
16. SILLMAN, S. & P. SAMSON. 1995. The impact of temperature on oxidant photochemistry in urban, polluted rural and remote environments. J. Geophys. Res. **100:** 11,497–11,508.
17. KATSOUYANNI, K., A. PANTAZAPOULOU, G. TOULOUMI, I. TSELEPIDAKI, K. MOUSTRIS, D. ASIMAKOPOULOS, G. POULOPOULOU & D. TRICHOPOLOUS. 1993. Evidence for interaction between air pollution and high temperature in the causation of excess mortality. Arch. Environ. Health **48:** 235–242.
18. LIPFERT, F. W. 1994. Air Pollution and Community Health. New York: Van Nostrand Reinhold.

19. MORBIDITY AND MORTALITY WEEKLY REPORT. 1993. **42:** 558–560.
20. MORBIDITY AND MORTALITY WEEKLY REPORT. 1994. **43:** 453–455.
21. SCHIMMEL, H. 1978. Evidence for possible health effects of ambient air pollution from time series analysis. Bull. N. Y. Acad. Med. No. **54:** 1052–1109.
22. KINNEY, P. L. & H. OZKAYNAK. 1991. Associations of daily mortality and air pollution in Los Angeles County. Environ. Res. **54:** 99–120.
23. LIPFERT, F. W. & T. HAMMERSTROM. 1992. Temporal patterns in air pollution and hospital admissions. Environ. Res. **59:** 374–399.
24. LI, Y. & H. D. ROTH. Daily mortality analysis by using different regression models in Philadelphia County, presented at the Colloquium on Particulate Air Pollution and Human Mortality and Morbidity, Irvine, CA, January 24–25, 1994. Inhalation Toxicology. In press.
25. SCHWARTZ, J. & D. W. DOCKER. 1992. Increased mortality in Philadelphia associated with daily air pollution concentrations. Am. Rev. Resp. Dis. **145:** 600–604.
26. WYZGA, R. E. & F. W. LIPFERT. 1994. Ozone and Daily Mortality: The Ramifications of Uncertainties and Interactions and Some Initial Regression Results, presented at the AWMA Specialty Conference on Tropospheric Ozone, Orlando, FL, May 1994. In press.
27. BURNETT, R. T., R. E. DALES, M. E. RAIZENNE, P. W. SUMMERS, G. R. ROBERTS, M. RAAD-YOUNG, T. DANN & J. BROOK. 1994. Effects of low ambient levels of ozone and sulfates on the frequency of respiratory admissions to Ontario hospitals. Environ. Res. **65:** 172–194.
28. THURSTON, G. D., K. ITO, C. G. HAYES, D. V. BATES & M. LIPPMANN. 1994. Respiratory hospital admissions and summertime haze air pollution in Toronto, Ontario: Consideration of the role of acid aerosols. Environ. Res. **65:** 271–290.
29. SCHWARTZ, J. 1994. Air pollution and hospital admissions for the elderly in Detroit. Michigan. Am. J. Respir. Crit. Care Med. **150:** 648–655.
30. THURSTON, G. D., K. ITO, P. L. KINNEY & M. LIPPMANN. 1992. A multiyear study of air pollution and respiratory hospital admissions in three New York State metropolitan areas for 1988 and 1989 summers. J. Expos. Anal. Environ. Epidemiol. **2:** 429–450.
31. ROGOT, E., P. D. SORLIE & E. BACKLUND. 1992. Air-conditioning and mortality in hot weather. Am. J. Epidemiol. **136:** 106–116.
32. MENDELL, N. R., F. W. LIPFERT, S. J. FINCH, L. A. BAXTER, R. GRIMSON, U. LARSEN, R. SINGLE, H. C. THODE, JR. & Q. YU. 1994. Ozone air pollution and human mortality: A review of studies applicable to estimating the external health costs of air pollution. Report AMS 94-13, State University of New York at Stony Brook, Stony Brook, NY.
33. DODGE, M. C. 1977. Combined use of modeling techniques and smog chamber data to derive ozone–precursor relationships. *In* International Conference on Photochemical Oxidant Pollution and Its Control: Proceedings, Vol. II. B. Dimitriades, Ed.: 881–889. EPA Report 600/3-77-001b, Research Triangle Park, NC.

[NOTE ADDED IN PROOF: More recent information (D. Hill, personal communication) suggests that we may have underestimated the current market penetration of residential air conditioning in New York City, which may now be closer to 80%. This would reduce the amount of new electrical generating capacity required and the concomitant emissions of air pollution. However, because so much of the New York City housing stock dates from the middle part of this century and before, much of the installed air conditioning capacity consists of room units, rather than central systems. It is not clear that cooling only part of a residence, say the bedroom, will appreciably reduce the penetration of polluted outdoor air into living space, so that health benefits could still ensue from increasing the amount of cooling per residence with respect to present levels by installing central systems and reducing the infiltration of outside air.]

Traffic and Transportation Planning
Strategies for Reducing Greenhouse Gas Emissions in the New York Metropolitan Area

JOHN C. FALCOCCHIO

*Department of Civil and Environmental Engineering
Polytechnic University
6 Metrotech Center, Room 515A
Brooklyn, New York 11201*

INTRODUCTION

For most people in the United States it is essential to own an automobile to participate in the job market, to go shopping, or to engage in a number of other activities.

The mass usage of the automobile has made possible the suburbs as we know them today. In suburban areas, as households, jobs, and income grow, auto ownership grows even more. And the result of high auto usage, without the ability to initiate corresponding increases in highway capacity, has created a condition where we can clearly see growth in traffic congestion without clearly seeing how we can effectively reduce it. Suburbs in the New York metropolitan area are no exception to this condition, although auto dependency is much less here than in other metropolitan areas.

In a similar vein, the truck has virtually replaced the rail car as a means to move freight in the New York metropolitan area. This heavy reliance on travel by motor vehicles whose fuel is either gasoline or diesel has created large amounts of greenhouse gas emissions that may be harmful to our future quality of life and health.

A review of the literature[1] indicates that motor vehicles are major sources of carbon dioxide (CO_2), carbon monoxide (CO), chlorofluorocarbons (CFCs) from air conditioners, and the precursors to both tropospheric ozone and acid rain: volatile organic compounds (VOCs) and nitrogen oxides (NO_x).

All of these gases contribute to greenhouse warming. However, CO_2 is by far the most significant greenhouse gas emitted by motor vehicles, in terms of the quantity emitted to the atmosphere. In the U.S., the light-duty vehicle fleet is responsible for about one-third of the CO_2 emissions. For this reason, the transportation planning strategies in this paper are illustrated in terms of their potential benefits in reducing CO_2 emissions in the New York metropolitan area. However, they also produce improvements in air quality through the reduction of emissions of carbon monoxide, volatile organic compounds, and nitrogen oxides.

HISTORICAL CONTEXT

Motor vehicle technology, together with its affordable prices, has shaped how land in the U.S. has been developed for the last fifty years.

Throughout history, urban settlement patterns have been shaped by the transportation technology available at the time. Thus the compactness of lower Manhattan or of medieval European cities and towns is the result of the need to keep distances as short as possible to maintain travel times within practicable limits.

As transportation technology advanced with the advent of the street car, followed by the subway and commuter rails, the boundaries of the crowded early settlements expanded as well, forming a pattern of urban activities concentrated in land parcels accessible by the transit routes. To maximize regional access to opportunities, the land developed in a pattern of high density and mixed use, and transit service and walking became very effective travel modes in meeting the mobility needs of urban residents. Today this type of urban form may be characterized as energy efficient, environmentally friendly, and ecologically sound, and it is often so advocated by those who seek to return to an earlier pattern of urban living when automobiles were yet to be mass produced. But as shown by recent trends, this is not to be.

Since 1950, most urban growth has occurred outside transit service routes; it is of the low-density type; it is dispersed throughout; and it is homogeneous as to use. This type of land use pattern also is a direct result of a new transportation technology: the automobile. The automobile is indeed the most effective travel mode to achieve maximum mobility in a low density, dispersed activity suburban system which is planned and designed without possibilities for walking or riding a bike to a destination and where transit service provides accessibility to only a small fraction of the total destination opportunities in the region.

This universal popularity of the automobile is made possible by a number of factors, including the realization that it is the most rational choice of mode to travel from place to place in suburban areas, and that the cost of traveling by car is quite low. It is paradoxical, however that the very popularity of the automobile is reducing its inherent attractiveness; the *collective increase* in car use in these areas is producing congestion levels that are having a negative impact on the effectiveness of the automobile in serving the mobility needs of suburban residents.

THE SOURCE OF THE CO_2 PROBLEM

Increasing reliance on motor vehicles in the movement of persons and freight results in increasing levels of air pollution and greenhouse gases. About 50% of the CO_2 in the atmosphere worldwide comes from fossil fuel. In the U.S., about 30% of CO_2 emissions from fossil fuel is from motor vehicles.[1]

It should be noted, however, that because of its extensive network of commuter rail, and its vast subway and bus transit system, the New York metropolitan area is the most energy efficient region in the U.S. on a per capita basis. For example, in this region 66% of all commuter travel is made by automobile, compared with 88% for the rest of the country. However, even with lower levels of auto use there is concern that the auto's share of the annual growth in personal travel in the future is *increasing*. New York City with its high-density development, its rich variety of land use mix where destination opportunities are within walking distance, and the most extensive transit system in the nation which carries approximately 25% of all transit trips made in the U.S. (ref. 2, Table 24) is the reason why the New York area is less dependent on the automobile.

In New York City, only 35% of commuter trips in 1990 were made by car.[3] But in the outlying counties, where most of the region's growth is expected, 85%

of suburban residents commute by car, a figure close to the national average.[4] Therefore, because much of the region's new development takes place in the outlying suburban counties, it is crucial that in these areas we focus on travel solutions that rely less on the use of the automobile and rely more on the use of alternative modes, including nontraditional transit, bicycle, and walking.

FACTORS IMPACTING ON THE AMOUNT OF CO_2 EMISSIONS FROM MOTOR VEHICLES

As shown in FIGURE 1, the amount of CO_2 emissions from motor vehicles is the result of three factors:

The Vehicle-miles of Travel (VMT)

This is the product of the number of vehicle trips and the distance traveled from one activity to another. (VMT may be expressed on an annual basis for each type of vehicle.)

The Energy Intensity of the Fuel Used

This variable, which determines how much fuel is needed to run a vehicle, depends also on the average travel speed prevailing at the time a trip is made. It is the inverse of vehicle fuel economy measure (miles per gallon). Fuel economy

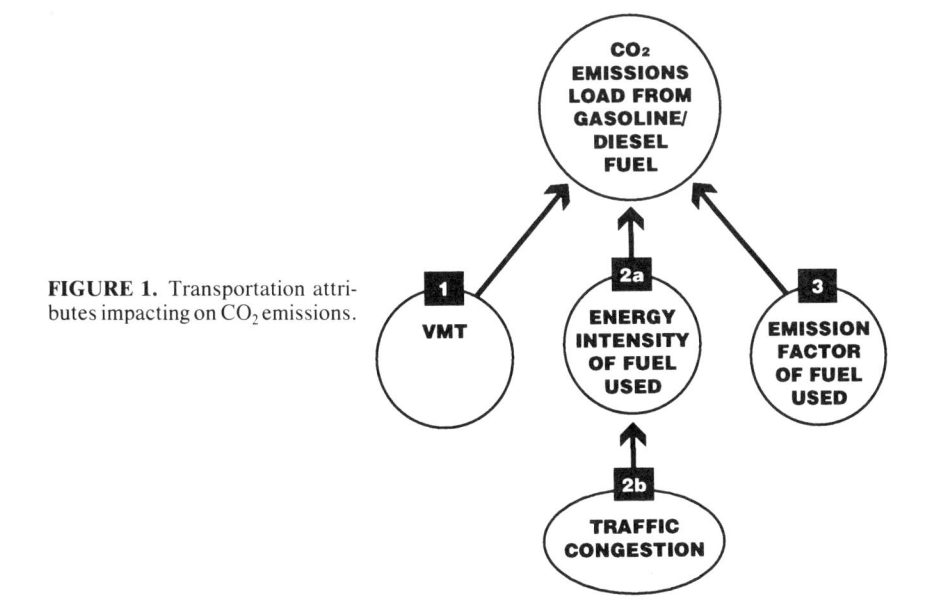

FIGURE 1. Transportation attributes impacting on CO_2 emissions.

is lowest in stop-and-go congested traffic, and highest for uninterrupted traffic speeds of approximately 50 miles per hour.

The Emission Factor of the Fuel Used

This variable represents all the CO_2 emissions associated with production, distribution, and use of the fuel. The current CO_2 emissions factor for gasoline is 23 lbs/gallon, or 86 grams/BTU (there are 125,000 BTUs/gallon).

The formula[5] to quantify CO_2 emissions may be expressed as:

$$CO_2 \text{ Emission} = \text{VMT} \times \text{Energy Intensity} \times \text{Emission Factor} \quad (1)$$

Thus, the amount of CO_2 emitted by motor vehicles can be reduced by reducing the values of any or all of the above variables.

HOW WE CAN REDUCE CO_2 EMISSIONS THROUGH TRAFFIC AND TRANSPORTATION

The following sections will describe how traffic and transportation planning strategies could be directed at reducing CO_2 emissions in the New York metropolitan area.

Strategies dealing with existing travel will yield the more immediate and more visible impacts, while those focused on how to better coordinate land development decisions with transportation objectives will provide a more fundamental structural fix to the root causes of excessive area-wide automobile use. But while land use/transportation coordination strategies are likely to yield more permanent benefits, these benefits will not accrue until many (50±) years in the future. FIGURE 2 illustrates how the desired results of reduced VMT and reduced congestion would be achieved by combining strategies dealing with new land development and strategies dealing with the management of existing travel and the management of the existing transportation system. The rest of this paper will describe how through transportation planning, we could reduce CO_2 emissions.

HOW WE CAN REDUCE DAILY VMT GROWTH

VMT growth in this region is the result of more people driving more frequently and for longer distances, and the result of increased truck usage.

The main reasons why people are driving more frequently and for longer distances are:

- An increase in population and jobs,
- An increasing disposable personal income,
- A dispersal of activities,
- Continued low-cost automobile travel which makes it easy to own and operate a car, and
- Lack of alternatives to the car in suburban areas.

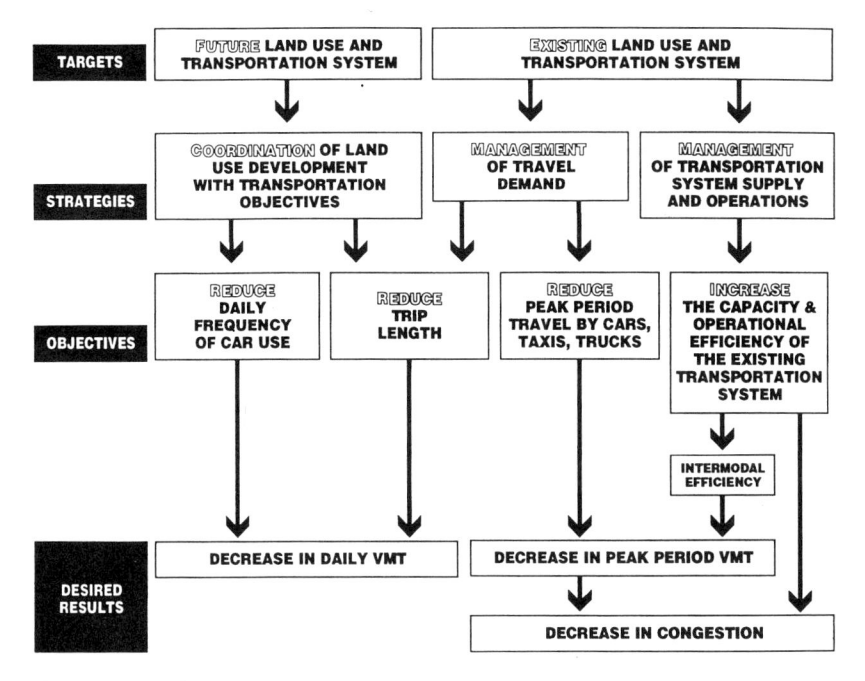

FIGURE 2. A rational approach to transportation planning for reducing regional VMT and traffic congestion.

In addition, more truck trips are occurring because we are entirely dependent on the truck to move freight across the Hudson River and to distribute it within the metropolitan area.

To reduce VMT growth in this region, we should focus on the following strategies:

- Coordinate land development with transportation objectives,
- Plan and design new developments and buildings such that access can be provided by transit, bicycling and walking,
- Improve transit services in mature urban areas,
- Improve intermodal connections between and among modes,
- Develop a bicycle path system to provide a new alternative to the use of automobile for short trips (less than 3 miles),
- Implement automobile trip reduction initiatives,
- Develop a network of HOV lanes, and
- Improve the rail freight system.

Coordinate Land Development with Transportation Objectives

This is perhaps the most difficult strategy to achieve, because to implement it requires changing how land development decisions and transportation decisions are made. For example, New York is a home rule state which means that how land is developed is the responsibility of cities, towns, and villages. Furthermore, these local jurisdictions are not obligated to develop comprehensive development plans. Thus, transportation improvements are being made and implemented by the State without requiring local levels of government to provide for land development patterns that minimize reliance on the automobile. Therefore, any travel time benefits provided by the newly added capacity are often quickly dissipated when the additional developments induced by these highway improvements are allowed to be built by the village, town, or city.

This condition creates a cycle that is the source of our congestion problems (FIG. 3). To solve the congestion problem in the long term is to break this cycle. Hence, as part of this region's transportation planning efforts, it is essential to

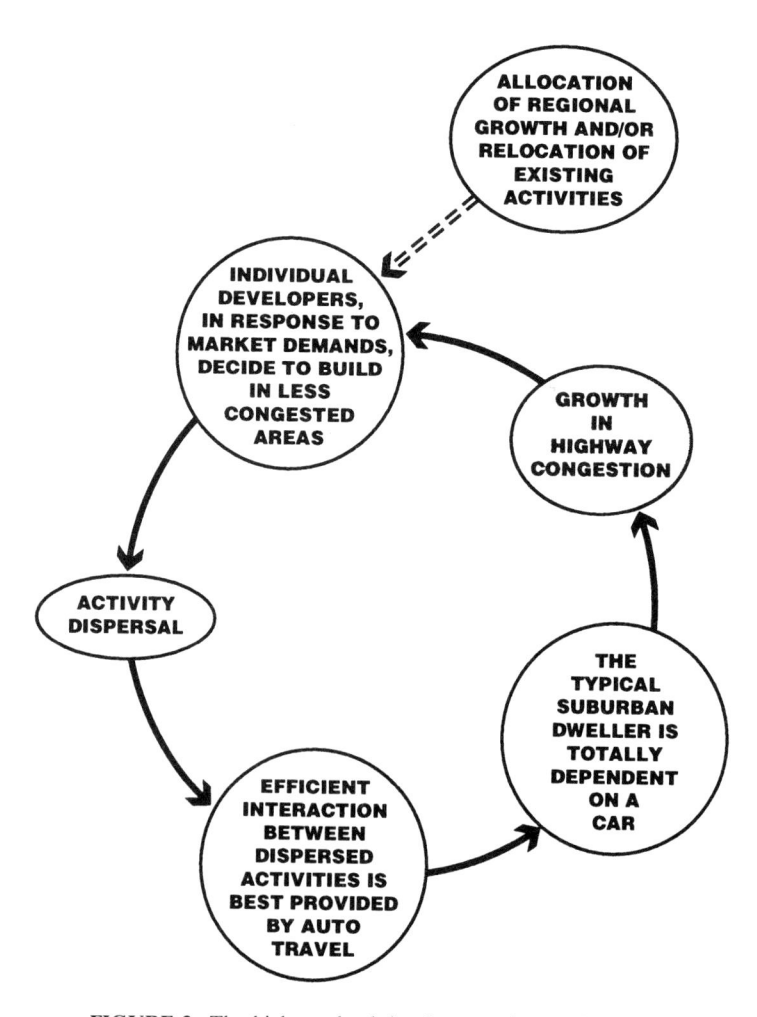

FIGURE 3. The highway-land development interaction cycle.

mandate a process that requires coordinating local jurisdictional decisions over land use with transportation objectives.

The following specific land use steps will lead us in the right direction for reducing VMT growth in this region:

- Encourage zoning changes to allow for mixed use development. The mixed use of the land will reduce the vehicle trip generation rates of individual land use if developed separately.
- The State should require municipalities to develop a land use plan which is consistent with the State's Transportation Plan.
- Encourage development where the transportation infrastructure capacity is available, or where the State has targeted corridors for transportation investments, and
- Discourage development where new transportation facilities must be built.
- Whenever a new public building is planned, State agencies should be required to give preference to downtown locations where alternatives to commuting by single occupant vehicle may be easier to find. As an example, the State of Oregon now requires state agencies to locate close to public transportation.[6]

Plan and Build New Developments with Access by Transit, Bicycles, and by Foot

Most projects built in suburban areas are exclusively focused on automobile access. We should provide other transportation alternatives to the car. For example, sports centers, shopping centers, and corporate parks should be required to provide for convenient and easy access by transit or biking. If these facilities were designed to provide for direct transit-access or bikeway access, there would be fewer car trips and more trips by transit and bicycle.

Improve Transit Services

Traditional and nontraditional transit services should be improved to serve especially the trip-chaining characteristic of the trip. For example, the aspects of trip chaining should be considered in pricing transit service. One possibility would be to sell rides by the expected duration of the trip. Thus, if a trip from work requires stopping at the cleaners, then the shoemaker, the picking up a child from the day-care center, and then go home, a person on such a trip should not have to pay multiple fares. If a ticket were issued for the expected duration of the total trip activity, say 90 minutes, then *one fare* would allow the flexibility of this type of travel, and could result in more people using transit and fewer people using the automobile.

Improve Intermodal Connections

By improving accessibility to transit services from other modes, and by interconnecting transit services more efficiently so that less effort is expended by travelers, we could expect a greater usage of such services and a corresponding lower usage of the car. A significant project currently under way in our region is the "Access to the Core" project being sponsored by the MTA and the PANYNJ.

Develop Bicycle Networks and Facilities

As discussed earlier, most suburban residents achieve mobility through the automobile. Those who use transit, whether they live in cities or suburban areas, do so either because they have no other choice or because their trip length by transit is less costly and/or less time consuming than if the auto were used.

The distribution of the lengths of trips made by persons in vehicles in a typical urban area will show that short trips of three miles or less represent approximately 50% of all trips made. This is a large market that conventional transit is not able to serve effectively. As a result, many such trips are made exclusively by automobile. The potential for using the bicycle to provide mobility for the short trip travel market should no longer be overlooked. Through the ISTEA legislation, the time has arrived to provide for safe bike travel and for facilities to park bicycles in a secure environment. In this region, we should rediscover the bicycle as a real alternative to the automobile for short trips involving either door-to-door travel or in getting to another mode, for example, to a commuter railroad station.

Encourage Implementation of Automobile Trip Reduction Initiatives

As it is currently practiced to meet the requirements of the Clean Air Act Amendments of 1990, this strategy involves working with employers of 100 employees or more to reduce the number of automobiles used by employees by requiring employers to develop plans to increase the average vehicle occupancy (AVO) by at least 25% over the existing levels in the area. Such a concept should be extended by State zoning laws to allow communities to offer incentives and bonuses in return for developer measures that reduce private car use through provisions that will facilitate bicycle and pedestrian traffic to and from the development.[7]

Develop Network of HOV Lanes

A very popular means of facilitating mobility in a congested network is through the development of highway lanes designated for use by higher occupancy vehicles. An areawide network of congestion-free HOV lanes would help reduce regional VMT, especially during congested periods, by inducing some drive-alone travelers to use a higher occupancy travel mode.

Improve the Rail Freight System

The New York City region has a greater percentage of freight moved by truck than the typical metropolitan area. It has been recognized by New York State that lack of adequate rail access for freight movement contributes to the high truck VMT in the region—especially for those highways connecting New York with areas east of the Hudson River. The State is funding improvements to the rail freight system such as elimination of clearance restriction on the Hudson line, construction of the Oak Point link, improvements of the Harlem River yard, the New York Cross Harbor project to revitalize car float operations, renewal of the freight operations of the Long Island Railroad, and greatly improved coordination between the Long Island Railroad and other railroad operations.[7] It is expected that these actions, by significantly improving the region's rail freight-services, will reduce truck VMT.

HOW WE CAN REDUCE CONGESTION

As was indicated earlier, energy consumption increases with congestion. An automobile moving at 50 mph on an urban expressway should use approximately an average of 0.04 gallons of fuel per mile. If the speed were to drop to 20 mph, owing to stop and go conditions, the gas consumption would increase to 0.10 gallons per mile. And if it were to fall to 10 mph, the gas consumption would increase further to 0.20 gallons/mile.[6] In fact, for congested traffic conditions small changes in speed in the order of 1 mph can change energy consumption in the order of 10% to 20% (ref. 8, TABLE 13.14, p. 475). Thus, it is apparent that any strategy that reduces congestion contributes significantly to the reduction of CO_2.

Traffic congestion can be mitigated by strategies and actions that:

- Reduce Peak Period VMT,
- Increase the capacity or operational efficiency of the existing highway system, and
- Prevent the deterioration of the traffic level of service of arterial highways in suburban areas.

Reduce Peak Period VMT

Strategies that reduce daily VMT are also likely to reduce peak period VMT. For example, if we implement new land development patterns that can satisfy personal mobility needs without requiring the use of an automobile for most cases, we also contribute to the reduction of peak period traffic congestion by limiting the number of peak period vehicle miles of travel. However, it should be noted that most of the future travel demand is from existing land use patterns. This implies that the type of actions needed to reduce the peak period VMT must induce current drivers not to drive during periods of peak demand.

Under severe traffic congestion, even a small reduction in VMT can be very effective in reducing congestion. For example, it has been noted that for nearly saturated traffic conditions on expressways, a VMT reduction of 3% can result in a 15% to 25% reduction in congestion.[7]

The type of actions supporting this peak period VMT reduction strategy include:

- peak period pricing
- telecommuting
- implement non-traditional transit service in suburban areas
- improving existing transit services

- encouraging employees to commute to work by sharing their vehicles with other drivers, or by using public transportation
- limiting taxi cruising in the Manhattan central business district
- parking subsidy reform

Those actions not described in earlier sections are briefly described below.

Peak Period Congestion Pricing

Of the techniques to reduce peak period VMT discussed in the literature, peak period congestion pricing seems to be the most cost-effective in theory.[9] The concept is based on the macroeconomic theory of supply and demand wherein the demand for a product or a service drops as the price rises. However, its implementation in the New York metropolitan area rests with the political process. Suffice it to say that in this region we have to move a long way from where we are to get to accept congestion pricing: commuters on toll facilities and in transit systems receive multiple pass/tickets discounts. So how are we going to be able to charge *more* than the average price when we are currently charging less than the average price? Obviously, we need to open up the topic for debate and the sooner we start the quicker we will get to a decision.

Telecommuting

Advances in telecommunications and their integration with computers and information systems makes the telecommuting option readily available to reduce or retime trips of some commuters. The New York region with long commuting distances and high congestion, should be ready to accept this concept. Employers of 100 or more employees may see this as a convenient element of their trip reduction program required by the Clean Air Act Amendment of 1990, and employees might perceive working at home a real benefit which reduces the stress they experience in their long commutes.

Implementation of Nontraditional Transit Services

In suburban areas, where densities are too low to support conventional transit service, it is essential to be innovative in providing alternative transportation services to the automobile. These would include a number of services that would effectively meet the mobility needs of commuters such as bus shuttles from railroad stations to employment destinations, and express van services along commuter corridors that could be driven in HOV lanes.

Limit Taxi Cruising in the Manhattan Central Business District

In New York City, there are 11,787 medallion taxis. The majority of the taxicabs is concentrated in Manhattan, south of 59th Street, where most fares are found. This is the Manhattan Central Business District (CBD) where the most severe congestion in the region is experienced.

It has been reported that about one half of the taxi mileage is spent searching for fares, while the other half is spent servicing passengers. Thus, 50% of taxi VMT in Manhattan could be substantially reduced by limiting cruising and establishing taxi stands where traffic congestion is highest. New York City has initiated a policy proposal for establishing taxi stand dispatching and ridesharing with the objective of streamlining taxi operations during the midday in the Manhattan CBD.[10]

Parking Subsidy Reform

Where appropriate, employers should charge employees for the free parking spaces they provide to those who drive. This concept would be an important tool in the implementation of employee trip reduction programs to promote ridesharing and transit use. In fact, "to facilitate the implementation of such a concept," President Clinton proposed changes to the Federal tax laws which would require employers who pay for employee parking to also offer a choice of an equivalent cash payment or tax-free transit pass. These cash out provisions seems to have worked well in California.[7]

Increase the Capacity and/or Efficiency of the Highway and Street Network

The peak period(s) traffic congestion on the region's highways and arterial streets is the result of four typical conditions:

1. Where the traffic demand on the roadway system, or corridor, exceeds the available capacity, or where the timing of traffic signals is not efficiently coordinated with traffic movement;
2. Where the illegal use of the traffic lane for loading/unloading blocks one or more lanes and restricts traffic flow;
3. Where incidents such as vehicle breakdowns and traffic accidents, especially on expressways and parkways, restrict traffic flow; and
4. Where it is made possible for property owners to access the arterial highway system without regard to the negative cumulative long-term effect of such decisions on the quality of traffic flow on the arterial highway.

Thus, to deal with highway traffic delay, it is essential that we approach the problem using a three-track strategy that includes the following:

1. Reduce recurring delay due to physical and operational deficiencies by providing additional capacity, improved traffic controls, and better management of curb space.
2. Reduce nonrecurring congestion, which accounts for up to 60% of the total delay for urban expressways, though the early detection and the early removal of incidents or accidents, and by providing travel information about transportation conditions to travelers not only while they are in transit but also before they start their trip.
3. The States of New York and Connecticut should accelerate their efforts towards the development of a State highway access code. Such a code, which is in effect in New Jersey, would greatly enhance the level of service of arterial highways in future years by developing comprehensive access management

programs specifying where and how land developments can have access to the arterial highways in the state. Such access management programs set forth driveways spacing requirements, minimum signal spacing, and the type of access (*e.g.*, all turns or only right turns permitted) allowed. The main concept behind the highway access code is to prevent strip development along arterial streets and to keep traffic conflicts resulting from the multitude of driveways and closely spaced signals from deteriorating the capacity of principal arterials located in relatively undeveloped areas.[11]

Achieving each of these strategies requires a commitment by Federal, State, and local agencies in working together to implement physical and operational improvements to the highway system as well as to implement "intelligent transportation systems" (ITS) to aid in the early detection of highway incidents so that they can be removed quicker and so that timely and accurate travel information is provided to the traveling public.

I am happy to note that for our region progress along the first two tracks is well underway as documented in the 2015 Long Range Plan by the New York Metropolitan Transportation Council.[12]

CONCLUSION

It has been shown that through transportation planning strategies and better land development planning and design, we could reduce greenhouse gases emissions from motor vehicles. This could be achieved through a reduction in VMT by better coordinating transportation and land use such that there will be less dependence on the automobile for mobility.

We can reduce greenhouse gases emissions by reducing peak period traffic congestion through several measures. Some of these measures are straightforward and deal with increasing the efficiency and capacity of the transportation system; others are more dependent on the political process for implementation (*e.g.*, peak period congestion pricing) and will require extensive public debate to determine if they will be feasible in this region in the near future.

Increasing traffic congestion was demonstrated to increase CO_2 emissions. It should be noted that traffic congestion also has negative effects upon air quality:

- Particulate emissions increase with congestion,
- Stop-and-go traffic on expressways increases tailpipe emissions of volatile organic compounds,
- Waiting for the green light at traffic signals also increases VOCs,
- Oxides of nitrogen emission are highest at low speeds.

Thus, mitigating congestion produces many other benefits besides that of reducing global warming. Some of the more obvious benefits of congestion reduction are the following:

- It increases economic efficiency by reducing travel time and the variance in travel time (*i.e.*, increases reliability of schedules),
- It reduces the number of accidents,
- It produces cleaner air,
- It improves health and reduces health costs,

- It reduces stress to travelers, and thus it tends to increase productivity at the workplace, and
- It reduces energy consumption.

If all of these benefits were aggregated and then compared with the cost of congestion reduction measures, there might be stronger support to justify the expenditures of additional resources to reduce traffic congestion and improve mobility. This concept is best expressed in the New York State Energy Plan:

> The integration of transportation, environmental, energy and economic development planning with congestion mitigation can be direct since common activities can be used to further the goals of each. The State should search for which common measures will best meet the multi-goals instead of offering a series of measures from different viewpoints which would work at cross purposes and waste scarce resources.[7]

In closing, it should be noted that greenhouse gas emissions will be reduced in the New York region by current plans by state and local agencies to improve transportation and air quality in compliance with regulations under the Intermodal Surface Transportation Efficiency Act of 1991 (ISTEA) and the Clean Air Act Amendments of 1990 (CAAA). However, additional actions that tend to reduce congestion but not VMT should be evaluated for implementation in cases where the net product is a reduced consumption of energy, and there is no negative air quality impact.

REFERENCES

1. WALSH, MICHAEL P. 1993. Highway vehicle activity trends and their Implications for global warming: The United States in an international context. *In* Transportation and Global Climate. D. L. Greene and D. J. Santini, Eds.: 1–50. Washington, DC and Berkeley, CA.
2. AMERICAN PUBLIC TRANSIT ASSOCIATION. 1992. Unlinked passenger trips by mode by transit system, fiscal year 1991(a). Transit Fact Book, 1992 edition. Washington DC.
3. PORT AUTHORITY OF NEW YORK AND NEW JERSEY. 1994. Home-to-work Flows by Mode of Travel—by County New York Metropolitan Region—29 Counties. March.
4. Journey-to-Work-Trends in the U.S., Based on Census Data. 1994. The Urban Transportation Monitor. Lawley Publications, P.O. Box 12300, VA 22009. June 10.
5. DE CICCO, JOHN M. 1992. The Greenwich machine: On the road to reduce CO emissions. Washington, DC and Berkeley, CA: American Council for an Energy-Efficient Economy. September.
6. Oregon Requires State Agencies to Locate Close to Public Transportation. 1994. The Urban Transportation Monitor. Lawley Publications, P.O. Box 12300, Virginia 22009. August 5.
7. NEW YORK STATE ENERGY OFFICE. 1994. Draft New York State Energy Plan, Volume II: Issue reports. February.
8. INSTITUTE OF TRANSPORTATION ENGINEERS. 1992. Transportation Planning Handbook. Englewood Cliffs, NJ: Prentice-Hall.
9. Twelve Tools for Improving Mobility and Managing Congestion. 1991. Urban Land Institute, 625 Indiana Avenue, N.W., Washington, DC 20004.
10. NEW YORK CITY DEVELOPMENT OF TRANSPORTATION AND DEPARTMENT OF ENVIRONMENTAL PROTECTION. 1992. Discussion Document—Traffic Congestion and Pollution Relief Study. May.
11. LEVINSON, H. S. 1994. Access management on suburban roads. Transportation Quarterly. Eno Transportation Foundation, Inc. Lansdowne, VA **48**(3): 315–325.
12. NEW YORK METROPOLITAN TRANSPORTATION COUNCIL. 1994. Critical issues—critical

choices: A mobility plan for the New York region through the year 2015. Draft. March 4.
13. WAGNER, F. A. 1980. Energy impacts of urban transportation improvements. Washington DC: Institute of Transportation Engineering.
14. GINSBERG, W. R. 1994. The Threat of Global Climate Change—What can New Yorkers do?: State and Local Strategies to Reduce Greenhouse Gas Emissions in New York State. Report of the Environmental Law Section of the New York Bar Association. January, p. 1.

DISCUSSION OF THE PAPER

QUESTION: I have seen estimates that about 10 percent of total vehicle-miles traveled are traveled while the person driving is lost. Certainly, in the typical trip between A and B, rarely do we go in the best possible route between the two. We've also seen in the last few years this new intelligent vehicle navigation technology coming about. Do you see this as having a potential role?

JOHN C. FALCOCCHIO: Oh, yes. In the paper I mention that. In fact in New York City, there are about 11 thousand, seven hundred something medallion taxis. Most of them operate within Manhattan, a very congested area. The way that they operate is such that half of the VMT that they put in is spent searching for a ride. Is that necessary? That could be reduced, too.

QUESTION: If 50 percent of the trips are three miles are less, what percent of the air pollution is due to these trips?

FALCOCCHIO: I would guess it depends upon the kind of pollutants that you are looking at. Those are the kind of trips that have a lot of cold starts. I don't know the numbers, but it could be substantial.

FREDERICK LIPFERT: I just want to come back to the air quality issue for a moment and reinforce what you said about land use. From my perspective, bicycles and cars do not mix in the same air mass. Anybody who exercises in the presence of automobile traffic doesn't understand the problem. Your breathing rate goes up, and you are taking in more stuff. Most of it isn't going to bother you, but who knows? You really need to get the heavy breathers away from the traffic, and in Manhattan you need also to think about getting rid of these internal combustion engines altogether.

FALCOCCHIO: I think that is being thought about. How far one is going towards implementation, I couldn't tell. But there is an awareness that what you say is a fact.

PHILIP JESSUP: Unless you're hyperventilating, you are breathing the same air as a bicyclist that you are breathing as a motorist or a pedestrian or a passenger.

"Growing" Energy-Efficient Physical Plants in the Greenhouse Era

LINDSAY AUDIN

*Energy Conservation Group
Columbia University
New York, New York 10027*

Buildings use energy for a variety of purposes, most of it derived from combustion of fossil fuels, either directly or through use of electric power generated by burning oil, gas or coal. Doing so adds to the carbon dioxide (CO_2) content of the atmosphere, which in turn heightens the greenhouse effect. Minimizing (or eliminating) use of fossil fuels, or switching to fuels that create less CO_2 per unit of released energy, therefore helps to control CO_2 emissions. At present, the main driving forces to make such changes are unrelated to climate change. Most are economic, being based on the costs of energy and energy efficiency measures (EEM). Such factors influence the cost-effectiveness of EEMs, as indicated by the payback period or return-on-investment (ROI) attained by reductions in energy use.

Taxation, regulation, and enforcement also affect the cost of energy and its attendant pollution, but are typically expensive to administer and limited in their impact. Utilizing market forces via pollution control credits and more efficient energy distribution and supply may, however, hasten reductions in emissions—without significantly disrupting the economy. Even the highest utility costs and shortest payback periods may, however, be insufficient to bring about major changes: dedicated capital and project management resources are still needed to make those changes actually happen.

END USES FOR ENERGY IN BUILDINGS

End uses for building energy are dominated by:

- space heating
- space cooling
- ventilation
- lighting
- domestic water heating
- vertical transportation (elevators and escalators)
- power-consuming processes (such as office equipment).

Other processes, such as cooking, washing, industrial pursuits, labs, *etc.*, are not considered part of the building and require specific measures tailored to each process. They will not be addressed here.

ENERGY SYSTEMS IN A TYPICAL CITY BUILDING

To place them into perspective, each EEM was applied to a typical ten-story, 100,000 ft² "glass tower" building in which electric, mechanical, and plumbing systems have remained unchanged since erection. Nearly 400 people work in the building and office applications are dominant. Personal computers are common, many of which are left on at night. The building uses .7 gal/ft^2/yr of oil to produce steam for heat and hot water, 24 kwh/ft^2/yr of electricity, and 3 ft^3/ft^2/yr of water.

Building Description

Glazing is sealed in place, and an asphalt roof covers an uninsulated concrete slab. The boiler uses #6 residual oil to produce steam for heating and hot water, and separate electric air conditioning systems serve each floor. Radiators heat perimeter spaces, and ducted constant volume reheat air systems separately cool interior and perimeter zones on each floor. Air temperatures are set by measuring return air flow, and radiator steam flow is controlled by an outdoor air temperature sensor. Fresh air flow is constant regardless of outside temperature. All heating, ventilation, and air conditioning (HVAC) systems remain available 24 hours a day. Spaces are lit by four-lamp, four-foot, fluorescent fixtures in a suspended ceiling. Wall switches are controlled by tenants. Standard motor-generator sets serve the elevators. Domestic hot water is fed from a steam-heated storage tank served by the building's boiler, and continually circulated throughout the building to ensure no delays in receiving hot water.

ENERGY-EFFICIENCY MEASURES AND THEIR IMPACTS[1]

Options are in approximate order of payback, adjusted for the necessary sequence of mechanical alterations. Many would be pursued simultaneously to reduce total cost and minimize disruptions.

1. *Install Power-Management Devices on Office Equipment.* Individual timer controls are added to reduce power wasted when printers and monitors (which often make up 60% or more of total computer wattage) remain unused for extended periods; computers and disk drives are unaffected. Similar devices are connected to copy machines. Payback is about 2 years, and yields an average reduction of .5 kwh/ft^2/yr.

2. *Upgrade Building Interior Lighting Efficiency.* Building light levels are examined for the tasks now being performed, and (where appropriate) lighting designers will lower them. Yellowed or inefficient lenses/diffusers are replaced with higher transmittance models, and lamp quantities are reduced to match new room needs. Fixtures are replaced with higher-efficiency models, or else improved through the use of specular reflectors. All standard cool-white fluorescent lamps are replaced by T8 tri-phosphor lamps, and magnetic ballasts replaced with electronic versions. In perimeter zones, daylight-sensitive dimming ballasts and controls are installed to reduce electric lighting when natural light is available. Most rooms receive occupancy sensors or timers that shut off lights when rooms are empty. Overall payback is about 3 years, and building power consumption is cut by about 6 kwh/ft^2/yr.

3. *Improve Domestic Hot Water Generation and Use.* Bathroom hot water use is reduced by adding flow restrictors to faucets and replacing showerheads with low-flow models. Losses in circulation piping are reduced by varying domestic hot water temperature with a programmable control that adjusts it according to patterns seen during a week's consumption. The hot water circulating pump motor

is replaced with a high-efficiency model, and a 7-day clock timer is added to shut it off when the building is generally unoccupied at night and on weekends. Overall payback for these measures is about 3 years, fuel use is reduced by about .01 gallons/ft^2/yr, water use drops by .5 gallons/ft^2/yr and total power consumption drops slightly.

4. *Improve HVAC Controls.* A building-wide energy management system (EMS) is installed, performing these functions:

- turns on/off all fans, based on operating schedules, and attains defined temperatures prior to occupancy each day
- modulates outside air dampers according to enthalpy, time schedules, and measured interior conditions
- controls A/C compressors according to cooling needs of individual spaces and occupancy schedules
- sequences heating and cooling so they do not occur simultaneously in the same zone
- adjusts space temperatures to minimize weekend and night energy usage, unless overridden by special tenant need
- controls cooling auxiliaries, such as pumps, to minimize usage when space conditions do not require heating or cooling.

Installation cuts steam consumption by about 25% and cooling energy by about 30%, yielding an overall payback of about 4 years. About 2.5 kwh/ft^2/yr and .15 gallons/ft^2/yr of fuel are saved.

5. *Increase Envelope Insulation and/or Solar Load Control.* The roof is insulated using urethane foam, cutting transmission losses and solar load for the top floor. Covering the single-glazed walls with low emissivity solar film increases its insulating value while reducing solar load. Overall payback for these envelope measures is about 5 years: fuel use drops by .02 gallons/ft^2/yr and electrical usage drops by about 1 kwh/ft^2/yr.

6. *Upgrade Elevator Motors and Controls.* The elevator's motor-generator set produces direct-current (DC) power (for a DC motor) from utility AC power. The generator idles when the elevators are not running, consuming power regardless of need. It is replaced by a silicon rectifier (SCR) controlled AC motor equipped with "sleeper" circuitry that turns off the motor when no call for elevator service is received for at least 5 minutes. Owing to the high cost of this conversion, payback from an energy standpoint is about 6 years and .5 kwh/ft^2/yr is saved.

7. *HVAC Design Improvements.* The reheat air systems (which warm cooling air to meet individual room requirements) are converted to variable air volume (VAV) units using inlet vanes at fans to adjust air flow as needed, instead of at the highest rate required on the hottest day. Steam radiators get thermostatic valves sensing individual room temperatures. These systems are controlled by additions to the EMS. Payback from purely an energy savings standpoint is about 7 years; .1 gal/ft^2/yr of fuel and 1.5 kwh/ft^2/yr of power are eliminated.

8. *More Efficient HVAC Equipment.* This upgrade is expensive, but is pursued in the future to replace aging equipment, increase building value, release floor space held by mechanical rooms, and attract higher-level tenants employing sensitive computer and telecommunications equipment (such as commodity traders).

Individual A/C compressor units on each floor are replaced by an efficient central chilled water system fed from a two-stage steam absorption machine (powered by the boiler plant) installed in the basement. The boiler's burner is replaced with a dual-fuel (oil/gas) burner, gas service is installed, and the boiler is downsized to better accommodate the building's reduced loads. Radiators are replaced by thermostatically controlled fan coil units that use either heated or chilled water. Fan motors are now controlled by variable speed drives (VSD) that minimize electrical usage better than inlet vanes. Building exhaust air passes through a heat recovery wheel that captures its warmth, or cools incoming fresh air. A central rooftop cooling tower serves the chiller plant, and uses VSD-controlled fans equipped with high-efficiency motors sequenced to minimize condenser water temperature when "free cooling" (from low outdoor temperatures) is available. The chilled water pump motor is controlled by a VSD, and the EMS is expanded to handle the upgrade.

Payback from a purely energy conservation standpoint is about 10 years, helped considerably by the reduction in fuel price when using dual fuels, and reductions in peak electric demand due to elimination of many small electric A/C compressors. Another 2 kwh/ft^2/yr is saved, but net fuel use (now also powering the steam absorption unit) *increases* by .2 gal/ft^2/yr of fuel.

9. *Improved Steam Distribution.* Most steam traps and condensate receiver tanks were eliminated when the steam radiators were replaced with hydronic (*i.e.*, water-based) heating and cooling. Thermostatic traps now reduce steam in the condensate lines, cutting piping heat losses. Steam piping is now concentrated in the basement mechanical spaces, serving only the boiler, fan coil hot water heat exchanger, absorption chiller, and domestic hot water heater. Energy payback is about 4 years once conversion to dual fuels has cut the average price of heating energy, saving .003 gal/ft^2/yr. Water use drops by .03 gallons/ft^2/yr owing to reduced condensate losses.

Net Effect of Energy Upgrades

Total electrical usage is reduced by up to 61%, while fuel usage drops by only 12% due to converting from electric cooling to steam-driven cooling. Using EPA figures[2] for the generation of greenhouse gases from regional power plants, this effort reduces CO_2 emissions by about 1000 tons per year for our 100,000 ft^2 building. Cost for all measures is $1–2 million dollars, depending on building conditions and how EEMs are pursued.

Options Not Implemented

- Double glazing (prohibitively expensive in a glass tower)
- Weatherstripping (not appropriate—windows not operable)
- Cogeneration (rarely cost-effective unless steam and electric loads are relatively coincident)
- Thermal storage (reduces electric bills but not energy)
- New electric chillers or gas-fired chillers/heaters (viable alternatives to the steam absorption system, depending on total installed cost)
- Renewable sources, such as solar and wind (not yet as cost-effective as the proposed efficiency items; while such sources could be installed once loads are lowered, even full elimination of the remaining consumption would not justify the present-day cost of renewable energy systems)

MINIMIZING EMISSIONS PER UNIT OF ENERGY

Greenhouse emissions can be further reduced by switching to fuels whose combustion emits less CO_2 per unit of released energy. To that end, these options stand out:

- Ending coal combustion in the City (which continues in many older facilities, including City schools)
- Extending gas and/or steam lines: buildings in areas not presently served are forced to use oil or coal-fired boilers
- Eliminating less efficient peaking units for power generation
- Using cogeneration facilities that wring more BTU's out of each fuel unit, such as combined-cycle gas turbines and fuel cells
- Purchasing power from more efficient utilities (and/or those using renewable resources) outside the City.

Such a variety of technical options would lead one to believe that our task should be easy. The trick lies in implementing them: for every good reason to conserve, this author has repeatedly heard ten reasons to do nothing. In the end, the viability of technical options is controlled by this balance of incentives and barriers.

INCENTIVES TO IMPLEMENTATION

Codes and regulations mandating implementation of EEMs have had limited success. Improved building codes (which primarily focus on new structures) do little, for example, in areas where growth and building turnover are relatively slow (such as New York City). Present codes (*i.e.*, New York State Energy Conservation Construction Code) do not even reflect the "state-of-the-shelf" in energy efficiency. Mere compliance thus assures that energy will be wasted immediately upon the opening of such facilities.

Economic forces provide most of the leverage for making energy efficiency happen. Five types of financial incentives have contributed to that end, all of them shortening payback period:

- Increased utility rates raise the price of energy
- Pollution regulations further increase the cost of energy
- State/federal grants lower the cost of installation
- Utility rebates cut the installed cost of equipment
- Pollution abatement credits reward cleaner energy use.

In the world of real estate, nothing speaks as loud as ROI.

BARRIERS TO IMPLEMENTATION

Few barriers are technical: most are historical, institutional, financial, educational, and/or managerial in nature. Such items surface whenever energy expenses are discussed by managers and owners, and must be kept uppermost in mind to avoid focusing purely on the technical solutions for reducing emissions.

1. *Energy Costs Are Not High Enough.* Despite heavy attention during the energy crises of the last 20 years, energy consumption by buildings has never

been a truly significant cost to New Yorkers: unless a homeowner is poor, direct energy costs (*i.e.*, that seen on a utility bill) rarely exceed 5% of a family's income. Many spend as much (or more) on monthly telephone and/or cable TV services as they do on energy. Buildings generating income (*e.g.*, rental real estate, commercial/industrial firms) rarely see energy bills greater than 2% of their annual revenues. Typical cost for personnel, for example, runs about $200 per square foot of office space per year—but all the energy used in service to that space rarely costs much more than $3 per square foot per year. Even the City of New York (which operates hospitals, schools, and municipal facilities) spends only about 1.5% of its budget on energy.[3]

2. *Older Building Stock Limits Impact of New Energy Codes.* Most of New York City was built during an era of cheap energy, and there is little prospect for demolishing such structures merely because they are not energy-efficient. Only a small portion of building square footage is replaced each year, so construction of new, more efficient, facilities is a slow process having only a minimal effect in the near term. Most energy codes apply primarily to new buildings (or changes to 50% or more of an existing building), so they rarely apply to typical renovations.

3. *Payback Periods Rule Every Energy Analysis.* Few EEMs pay for themselves instantly; most take several years to do so, and often longer than indicated by often simplistic analyses. Many owners choose not to invest beyond a given ROI, or what regulations require. Competition for capital (for a new motor or a new disco) says that money goes where it yields the quickest return.

4. *Energy Pricing Is Often Immune to Society's Needs.* While there is competition among some energy suppliers, energy price is often unrelated to energy supply or demand. Instead, it is strongly influenced by costs of development, generation, and other factors. Various layers of regulation have created a maze of processes that influence prices, many of which bear no relation to the long-term impacts of energy use. Cogeneration, for example, is seen as a threat by utilities, so they pursue every avenue to block it, including charging very high backup rates that discourage such installations.

The EEMs that are most cost-effective under such rates are often not those that most reduce CO_2 emissions. Con Edison's electric rates are designed to recover costs for power plants and distribution systems, resulting in very high rates for the monthly peak power draw (measured in kilowatts) and relatively low rates for electrical consumption (measured in kilowatt-hours) rates for large consumers. As a result, the payback periods for many options (such as lighting controls) that cut energy use but not many peak kilowatts may be too long to attract investment. The steam absorption chiller in our upgraded building, for example, provided the best overall ROI, due primarily to its avoidance of high peak demand charges. It may, however, cause release of more CO_2 per unit of cooling than an electric chiller.

5. *Energy Distribution Systems Are Not Coordinated with Needs.* New York City is interlaced with steam, gas and power lines that do not fit into any overall energy efficiency or pollution control scheme. Utility steam, for example, is only available in Manhattan up to 96th St., and high-pressure gas stops at 137th St. Both limits force use of boilers firing #6 oil or coal, resulting in more CO_2 per unit of used energy than the other sources.

6. *Building "Balance Sheets" Are Uncoordinated.* Building operations are divided among departments with competing agendas that derail plans for efficiency. Buying efficient components, for example, may be blocked by a purchasing agent working with a limited budget, or one seeking kudos for cutting expenses.

7. *Real Estate Economics Limits Opportunities for Efficiency.* Those involved in developing and using City property often choose to forego energy improvements. Unfortunately, there is no law requiring that waste be avoided: our economic system leaves that choice up to the property owner.

- Landlords pass energy costs along in rents, so they are not seen as an expense needing attention (unless the ROI of an EEM is better than other investments). Many lack knowledge to cut their energy use, or had bad experiences with past efforts.
- Renters typically have little or no control over heating systems, do not own many of the larger appliances they use, lack the right or ability to alter the electrical or mechanical systems serving them, and would be unable to recover investments in building efficiency once they leave the premises.
- Homeowners (or co-op dwellers) have wider options, incentives, and abilities, but often lack the funds for EEMs unless they are pursued as part of an otherwise necessary upgrade (such as window or roof repairs).
- Builders gain nothing by spending extra money on efficiency—unless required to do so by code, or an owner (with a great deal of foresight) who also plans to occupy the new building. Elective EEMs that increase building costs are the first to be dropped to cover cost overruns during construction.

8. *Control of Building Energy Usage Is Quite Diffuse.* A relatively few large users consume as much as hundreds of thousands of smaller users. Most electrical and natural gas accounts, for example, are for single-family residences (either apartments or houses) where consumption is dominated by a few appliances (*e.g.*, boilers, hot water heaters, refrigerators). Larger users, on the other hand, have operating staffs, financial goals, and heavy investments in their physical plants. Fostering conservation thus requires a variety of tactics and incentives.

9. *Facilities Management Is Traditionally Low-Tech.* While the technologies of energy management have advanced considerably over the last 20 years, those in the position to use them are notoriously behind the times. "Blue collar" thinking and attitudes are common where "lab coat" perceptions are needed instead. While training is available, few facilities provide career development to that end.

10. *Many Energy Policies Are Outdated and Myopic.* Through a variety of subsidies, incentives, and misguided efforts, government and energy suppliers have created the notion that "more is good, and cheaper is better." Instead of, for example, providing funding tied to improving energy efficiency (and thus permanently cutting costs), special electric rate *reductions* are given to large energy users to keep them in the City. Energy efficiency is thus subverted to other goals (jobs and economic development). But "installing a more efficient light bulb, air conditioner, or motor in a factory can reduce your electric bill just as surely as a rate reduction."[4]

MOVING TOWARD SOLUTIONS

Define a Higher Value for Clean Air

At the heart of many of these problems is a missing component: a clearly-defined price for acceptably clean air, free of CO_2 and other major pollutants. Were such a price established, other barriers would be minimized as existing market forces came into play to reduce the expenses associated with it. If, for example, the average cost of fossil fuels was raised by society's cost to mitigate the effects of their associated pollutants, the payback periods for pollution abatement and energy efficiency equipment would drop, making them better investments. Once offered as a "carbon tax" during the early discussions of energy in the Clinton administration, such a fee would selectively target the most-polluting fuels and least efficient systems. The cost of renewable energy sources that do not create CO_2 would, however, remain the same—making them a relatively better investment.

For example, power plants that discard most of their energy through cooling towers (instead of cogenerating steam or hot water) would cost more to operate, making them uncompetitive with other generators until re-designed. (This concept presupposes, however, that utility competition eventually becomes a reality.) Fuels and energy-conversion processes (such as natural gas and fuel cells) that emit less pollution per unit of usable energy would get a boost. Markets for pollution abatement credits would develop more rapidly, encouraging greater efficiency wherever feasible while allowing orderly phaseout of older systems. At present, such markets are hampered by the slow development of rules and prices, and the relatively low cost of the most polluting fossil fuels.

Alternatively, "clean air pricing" could be levied to increase the cost of the least efficient end-user devices. Incandescent bulbs, for example, are one of the cheapest—and least efficient—devices in existence. Translating the clean air price into a flat "watt tax" would raise the cost of incandescents much more than their compact fluorescent (CFL) competitors, equalizing their first costs, and eventually making incandescent bulbs as obsolete as whale oil lamps.

Develop Sound Energy Policies and Regulations

These goals can be further aided by governmental action, such as:

- Improving efficiency specifications, building codes, and construction standards, and making them applicable to *all renovations* exceeding a low area limit (*e.g.*, 5,000 ft^2).
- Converting rate abatement programs for jobs and business into energy efficiency programs achieving the same results.
- Eliminating exemptions for City-owned facilities still burning coal, and installing less-polluting systems.
- Discouraging (through regulation, taxation, and standard-setting) the most wasteful/polluting sources and end uses of energy; *e.g.*, coal burning, electric heat, incandescent lighting, and unnecessary conditioning of outside air.
- Pressing for true competition among utilities so that low-cost clean energy sources become accessible.

Expand the Supply of Qualified Technical and Management Personnel

To take full advantage of the changes proposed above, enhance the technical expertise of those involved in implementing EEMs by:

- Improving and tightening professional certification and training/education of those designing and specifying energy-using systems and equipment.

- Funding and deploying experienced energy "SWAT teams" that help control energy use without interfering with operations.
- Creating and supporting regional energy/pollution managers that maintain pressure on municipalities and large users to cut energy use through routine site visits and monitoring.
- Funding inspector training in existing/future building codes, and consistently enforcing them.

Foster Availability of Less-Polluting Energy Resources

Some efforts cross over the boundaries between policy-making and energy marketing, and include:

- Eliminating barriers (such as resistance to open transmission access, also called "wheeling") to buying the least-polluting power.
- Encouraging markets in pollution abatement credits by gradually tightening emissions allowances via action at state/local level (similar to auto emissions compliance efforts).
- Restoring the Institutional Conservation Program (run by the State, using federal oil overcharge funds) that funded energy upgrades in state/municipal/non-profit facilities in the City.
- Expanding/restoring Con Edison & New York Power Authority energy rebate/loan programs to "move the market" toward more efficient equipment.
- Eliminating utility barriers in rebate programs to finance systems that save energy, even if they also generate power.
- Fostering replacement of oil and coal-burning by pushing (and, if necessary, mandating) extension of high-pressure gas and utility steam lines and services to more of the City.

By pursuing these options, we can greatly accelerate progress toward safeguarding our atmosphere.

REFERENCES

1. Kapadia Consulting, Inc. 1994. Technical Assistance Audit of 570 Lexington Avenue. Peekskill, N.Y.
2. USEPA Green Lights Electronic Bulletin Board. 1994. Washington, D.C.
3. Personal discussions with Katharine Luthin. 1993. NYC Office of Energy Conservation. New York, N.Y.
4. Freeman, S. D. 1994. A Report to the New York Power Authority Trustees on NYPA's Restructuring and Cost-reduction Review. New York, N.Y.

APPENDIX ON WATER CONSERVATION

Atmospheric heating due to the greenhouse effect could increase demand for water while decreasing its availability, so efficiencies in water usage may become an essential part of policies to cope with it. Process loads (cooking, industrial labs, *etc.*) require individual attention and are not considered part of building use.

Building Water Usage

Buildings typically consume water for:

- waste handling (toilet flushing, process loads)
- washing (sinks, showers, floors, *etc.*)
- cooling (evaporative cooling towers, once-through systems)
- heating (steam systems)
- elective uses (*e.g.*, swimming pools, fountains).

Water leakage is also a significant "hidden" consumer of water.

Options for Water Conservation

1. Since *toilet flushing* is the largest consumer of water in buildings (typically 60–75% of total consumption), replacing the standard 5 gallon per flush (GPF) models with modern 1.6 GPF toilets significantly reduces building water use. In 1994, New York City initiated a program giving rebates to building owners for replacing toilets.

2. Adding flow restrictors to *sink faucets and showerheads*, when done properly, can cut their usage by more than 50%. The City's toilet program requires that low-flow models replace standard showerheads in bathrooms receiving new toilets.

3. Once-through *cooling systems* (common in labs and computer spaces experiencing increased equipment loads) use water to absorb A/C compressor heat, and then dump it into the sewer. Eliminating such systems (already restricted to 2 tons or less of cooling), or recycling the water for toilet flushing, would reduce use by about 1.5 gallons per minute (GPM) per ton of cooling, which translates to nearly 100,000 ft^3, a year for a continually running 1 ton unit. Feeding such units with chilled water from central cooling systems would cut water waste without greatly increasing cooling energy use.

Mechanical cooling often uses *evaporative condensers* called "cooling towers." Shutting off unused lighting, computers, and office equipment allows reduction or shutdown of night cooling, including such towers, which typically evaporate 5% of their water flow. Unneeded night cooling in our example building could waste up to 300,000 cubic feet of water each cooling season.

4. Reducing losses in *steam heating systems* (common in most buildings having their own boilers) saves both energy and water. When steam condenses, returning the treated "condensate" saves about 10% of the energy needed to re-vaporize it. In older systems, condensate losses (from leaks and failed equipment) of 20% are not unusual. In our example building, such losses could exceed 27,000 ft^3 of water (and over 1,000 gallons of heating oil).

5. *Elective systems* losses are controllable through timed pumping and after-hours covers for open pools and fountains. Chemical treatment plants employ automated pool covers, or use floating plastic balls to cut evaporation. The balls are released into the pool, float to the top and cut evaporation by reducing the open surface area of the pool or fountain; they are automatically removed and held in a catch net when not in use.

6. *Recycling "gray" water* (from sinks and showers) into toilet flushing is possible, and is enhanced when mixed with collected rainwater. Some filtration or treatment may be needed to control odor and scale, but using such non-potable sources reduces the need for City water. Water storage becomes necessary, but may already exist in "house tanks" on building roofs, presently used to maintain domestic water pressure. Alterations to existing plumbing would be necessary, but piping used for other purposes (such as roof drainage) could become the new conduit for sink and shower output, allowing the "black" water from toilets and urinals to continue using the existing sewage system. A separate fresh water supply for potable uses would also be needed, but employing low-flow toilets, faucets and showerheads would greatly reduce its cost and pipe size.

7. Surveys of City housing[1] found that *fixing plumbing leaks* in toilets, faucets, and piping was often more cost-effective than installing flow-restrictive equipment, and the two tasks should be pursued in tandem. For example, a leaking standard toilet valve can waste up to 2 GPM between flushes, which (in a year's time) is over 100,000 ft^3 *per toilet*. Even a trickling faucet can waste .1 GPM (about 1 cup per minute), or about 5,000 ft^3, per year *per faucet*.

Factors Limiting Waste Reduction

As with energy efficiency, most limits to implementation are non-technical, but since water services are supplied by the City government, one must now add "political" to that list of limitations.

- *Pricing:* Fresh potable water is grossly underpriced, and not even fully metered, so its waste does not catch much attention.
- *Payback Period:* Cost for plumbing upgrades is relatively high, so the low dollar value of saved water yields payback periods too long to attract private investment capital.
- *Political Will:* Because of the impact on low-income tenants in water-wasteful buildings, water metering has not been popular.
- *Water Billing:* City billing is irregular (at best), making it a nuisance and difficult to monitor both usage and savings.
- *Past Quality of Water-Conservation Equipment:* Some devices do not work well on older systems (*e.g.*, low-flow toilets), and require extra care in their specification and purchasing.
- *Poor Metering:* Inappropriate metering locations often obscure sources of waste (*e.g.*, one meter covering several buildings). City meters lack the capacity to be read electronically, severely limiting their usage for leak detection and flow monitoring.
- *Government Priorities:* The City's focus has been on collections for water and sewer usage, not waste reduction. When end-users bear these costs, City motivation for waste reduction disappears.
- *Poor Building Codes:* Buildings are not designed for efficient water use because the low prices for water and sewerage provide no incentive to spend extra money on water-conserving equipment.

Present Motivations for Water Conservation

Increased environmental regulations require the City to improve its sewage handling abilities, or else reduce the need for them (which means reducing the amount of water dumped into sewers). The unit cost to expand system capacity is now greater than the unit cost to decrease end use demand for sewer services. The City's toilet program is a direct result of that realization.

Pressure for conservation increased during several recent summmer droughts as City reservoirs reached alarmingly low levels. Emergency declarations mandated facilities to cut back, but usage climbed again when the alerts ended.

When the price of water was rising almost 9% per year, some building managers took note. But further increases were frozen when some tenements could not pay their water bills, taking any pressure off managers to change their wasteful ways.

MOVING TOWARD SOLUTIONS

Making water conservation happen requires the same focused attention (and many of the same methods) as needed for energy efficiency. Some of these measures could also have immediate salutary impacts toward avoiding future droughts, controlling housing costs, and stretching the lifetime of the City's water system.

REFERENCE

1. HARMON, R. 1992. Studies of Low-Income Housing by the Brooklyn Ecumenical Cooperative. Brooklyn, NY.

DISCUSSION OF THE PAPER

QUESTION: You said that retail wheeling could be designed so that it would impact price in a positive way. Could you elaborate on that?

LINDSAY AUDIN: Yes, I'm glad you asked the question. Let me give you an example of exactly how this would occur. A number of years ago, natural gas was deregulated. I buy natural gas on the spot market month by month. I buy it, actually, from the Gulf of Mexico. Then I have it piped up here, and I pay a transportation charge. New York State got real smart, and they said, "Gee, you're avoiding all the taxes, because this is now an interstate piece of commerce. We are going to slap a tax on you that is going to be called 'the gas importer's tax.'"

Now, this gas importer's tax took away some of the savings I got from buying my gas in a cheaper place, and the state took it and put it toward general revenue. Retail wheeling is a very similar situation. If I buy my electricity from another source where it is cheaper, I would hope—because the difference in cost is enormous, it's twenty to forty percent cheaper—that the State of New York would get smart and say, "Gee, I think I'll tax that at the border, grab a chunk of the money, and use it to refund the institutional conservation program, or utility rebate programs," or whatever. But allocate the savings that occur from retail wheeling, a portion of it, directly to energy conservation.

That's Step One. Step Two. When you get to the situation where prices do drop, or where there are economic development opportunities, some of that money is allocated, through taxation or another type of regulatory method, but you only get it back if you use the money to do energy conservation work. So, for example, if I get an economic development package because I built a building, or whatever, the State of New York says, "Fine, you get thirty percent off on your electricity for the next ten years, but you are only going to get all that thirty percent if you spend half of it for energy conservation measures."

Now, I'll spend that money because I still want that other fifteen percent. In the end, I know I am going to get the same result, because the other fifteen percent that I spent on energy is going to get me more than that back in reductions over time. And once that abatement goes away, I'll still have the reductions in my bills, because I will now have done it through conservation, not just politics.

QUESTION: I just want a clarification. In your talk, you said that renters cannot do anything about heating costs because of central operations.

AUDIN: The biggest chunk that you've got is commercial real estate as far as consumption goes. But in residential real estate, there are some opportunities, depending upon who owns the building. It has been my experience in dealing with large apartment buildings that if someone calls up and wants the heat reduced, they are probably the only person in the building who wants the heat reduced. As a former apartment renter, I know that if I ask to have the heat reduced, it will probably stay reduced for the rest of the wintertime. Sure, you can call up and ask for the temperature to be reduced if it is too high, but the way that most people control things is with a thermostat made of glass. It's called a window. They open the window. And that's how most people control their energy usage. Now, most large buildings do not have thermostats in the rooms. They have something called a heat timer, a device that measures the outdoor temperature, the indoor temperature at one point, and cycles the boiler off and on based upon the difference. That is completely insensitive to an awful lot of the variables in the building. If the building is zoned—north and south, east and west, whatever— the heat timer will turn on heat for the coldest spot in the building. Everybody else will be overheated, and they will open the window. So even if you called to have the temperature reduced, one of the ways that might occur is an adjustment to the heat timer. Whereupon, if you are on the west side of building, the people on the north side will be cold and will call back and say, "Raise it back up again." So in many cases, whatever you do in a building is greatly dependent upon its control system.

WAYNE TUSA: Any comments on EPA's Green Lights program?

AUDIN: As a matter of fact, yes. I was the first to bring a university into the Green Lights program. One little simple point on that, when I first introduced EPA to the concept of universities in 1991, their answer was: "Universities? You're nonprofit. Why do you want to join this program?" We had to twist their arm to get into the program. We were one of the winners in the Green Lights achievement awards last year. We were beaten out only by Mobil Oil Corporation, which happened to operate faster and have more money to spend than we did; otherwise, we would have been first. Green Lights is a very good start.

But Green Lights is focused on lighting. There are other efforts that I went into in my paper that I won't go into now. There is now an Energy Star program to try to foster more efficient computers, PCs, printers, *et cetera*. Those are all good programs being pushed by EPA. If Clinton had not been elected, the Green Lights program would be at about a quarter of where it is right now. It started under Bush, but when Clinton got in he tripled the budget for the Green Lights program. Now it is moving ahead like a rocket. It has literally thousands of large corporations, institutions, and even a few states signed on to it. The bottom line of Green Lights is: you agree within five years to upgrade your lighting to certain standards set by EPA, which are much better than any national or local or state code. Still not as good as it could be, as far as I am concerned, but if everybody met Green Lights it would have an enormous impact on the electricity that is being wasted right now running things like incandescent lights. I think it is a really good idea.

Energy Demand and Supply in Metropolitan New York with Global Climate Change

SAMUEL C. MORRIS III AND GARY A. GOLDSTEIN

Biomedical and Environmental Assessment Group
Brookhaven National Laboratory
Upton, New York 11973-5000

A. SANGHI

New York State Energy Office
Two Rockefeller Plaza
Albany, New York 12223

DOUGLAS HILL

Douglas Hill, P.E., P.C.
15 Anthony Court
Huntington, New York 11743-1327

INTRODUCTION

The purpose of this paper is to provide background information for a scenario planning exercise. It draws on several different sources, but the primary ones are an evaluation of carbon dioxide emission control strategies using the energy and environmental systems model, MARKAL,[1] and the New York State Energy Plan, its technical appendices and other reports from the State Energy Office.[2-5] The former applied computer modeling that is sensitive to cost assumptions. The modeling period extended to 2028. The analysis was intended to explore the ranges of possibilities rather than to assess policies that were actually under consideration by the State. The State Energy Plan draws on a wider range of information and analysis techniques. It provides projections to 2012, but often focuses on current action and proposals. The two are not in conflict; they serve complementary purposes.

A different consideration is that the sources address the energy system of the State of New York as a whole, whereas here we are concerned with the metropolitan area, which includes just the southernmost portion of the State and parts of Connecticut and New Jersey. We will try to extrapolate conclusions from the statewide data to the metropolitan area, in part through the use of a revision of MARKAL that distinguishes between upstate and downstate energy demand and supply.

When we talk about energy systems planning we focus on what should be done in the next few years, with a planning horizon of, perhaps, 30 or 40 years into the future. To consider potential climate change in that planning, we must be aware that decisions made now may affect a much longer time horizon, but our focus on the near term does not change.

EFFECTS OF CLIMATE CHANGE ON ENERGY SUPPLY AND DEMAND

Global climate change—which we assume for the Northeast United States means increasing temperatures, rising sea level and increasing severity of storms—would affect the supply and demand of energy in the New York metropolitan area in several ways.

Global warming may reduce the volume of winter snow pack. This would lead to increased winter runoff and reduced spring and summer runoff.[6] Warming is also expected to bring a decrease in summer precipitation to the Northeast U.S.[7] Both result in reduced hydroelectric generation in the summer when it is most needed. This would have a substantial effect on New York State, decreasing its own production as well as its potential to import hydroelectric power from Canada. The metropolitan area currently does not receive much hydroelectric power. As shown in TABLE 1, only 3 percent of the inexpensive power from the New York Power Authority's Niagara and St. Lawrence River-FDR hydroelectric plants reaches downstate. (Downstate is defined here as essentially the service areas of Con Edison, Long Island Lighting Company, Orange & Rockland, and Central Hudson Gas & Electric.)

Warming, *per se*, would have little or no impact on the supply of fossil fuel energy the Metropolitan area now depends on. If the temperature of water bodies increased, the efficiency of steam-electric plants would reduce slightly because it is dependent on the difference in temperature between the steam generator and the cooling water.

Increased temperatures will increase the demand for air conditioning. One study estimated that a 3°C increase in mean temperature might increase net annual electricity use by 2% because the corresponding decrease in winter heating demand would not offset the larger increase in air conditioning demand.[6] The effect on the summer peak would be greater; the increase was estimated to range from 2.9 to 6.7%. The downstate utilities have peak loads in summer, ranging up to—in

TABLE 1. New York Power Authority Sales to Customers in 1993 (thousands of megawatt-hours)

	Upstate	Downstate	Out of State	Total
Hydroelectric				
Niagara	10,881	519	1,819	13,219
St. Lawrence-FDR	5,634	41	537	6,213
Total Hydro	16,516	560	2,356	19,432
Percentage of total	85	3	12	100
Percentage in-state	97	3	0	100
Other				
Fitzpatrick	1,540	3,687	0	5,228
Poletti/Indian Pt 3	507	8,791	63	9,361
Marcy	74	397	44	514
Blenheim-Gilboa	5	19	0	24
Grand total	18,641	13,454	2,463	34,558
Percentage of total	54	39	7	100
Percentage in-state	58	42	0	100

SOURCE: New York Power Authority, 1993 Annual Report, pp. 40–43.[10]

TABLE 2. Apportionment of Energy Demands between Downstate and Upstate in the MARKAL Model of New York State

Demand	Downstate	Upstate
Commercial		
Space heat	52%	48%
Water heat	62	38
Air conditioning & ventilation	76	24
Lighting & appliances	66	34
Miscellaneous use	70	30
Industrial	26	74
Residential		
Space heat	60	40
Water heat	67	33
Air conditioning	75	25
Cooking	75	25
Miscellaneous appliances	62	38
Transportation		
Automobile	53	47

SOURCES: Maynard Bowman, New York State Energy Office, Projected energy sales by utility service area, personal communication, September 1994. Nathan Erlbaum, Data Service Bureau, NYS Department of Transportation, personal communication, 22 September 1994.

the case of Con Edison in 1993—40 percent higher than their winter peaks. The upstate utilities as a group, on the other hand, have a winter peak load, in total about 6 percent higher than their summer peaks (ref. 8, pp. 14–15).[8] This means that in the downstate area with summer peak loads, there would be a greater need for more megawatts of generation capacity to meet peak loads as well as the need for more kilowatt-hours of electricity.

The difference in seasonal peak loads between upstate and downstate was achieved in the model by separating the energy demands projected for the two regions, as shown in TABLE 2.

POLICY TRENDS AFFECTING ENERGY SUPPLY AND DEMAND

Anticipation of global warming whether or not it is a reality, may lead to policies that discourage the use of fossil fuels. The New York State Energy Plan (NYSEO,[2] p. 141) suggests that "increasing concerns with global warming and carbon loading in the atmosphere could severely constrain the use of coal. In 1992, coal provided over one-fifth of New York's electricity . . ." Oil produces 60–70% of the carbon emissions of coal and would also be at risk from such policies. Increasing emphasis may be expected on renewable energy sources and on energy conservation. Even without policies based on global warming considerations, current policies to reduce more conventional pollutants and toxic pollutants under the Federal Clean Air Act lead generally toward increased conservation, renewables, and natural gas, although "end of the pipe" emission controls on sulfur and nitrogen oxide emissions often result in higher carbon emissions.

UNCERTAINTIES TO BE CONSIDERED IN PLANNING ENERGY SUPPLY AND DEMAND

The following are questions that cannot be answered, but that have a significant impact on planning for a future energy system.

Future growth in energy demands. Energy demand growth will depend in part on improvements in efficiency of motor vehicles, air conditioning units, *etc.*, but will probably be influenced more by behavioral changes in society. Will more people drive longer distances to work? Will suburban life shift to larger houses with fewer people in each? Will income rise, leading more people to use more energy for air conditioning, hot tubs, motor boats, *etc.*? Increasing or decreasing energy demands will make the greenhouse problem more difficult or less difficult to solve.

Future energy prices. Energy prices affect demand. If the price of oil increased drastically, for example, economics will drive decreases in automobile use. The metropolitan area suffers from high electric rates. These impose a burden on the public and on industry, but they do help to keep demand down. A proposal for the New York Power Authority to take over the Long Island Lighting Company is only one of many proposals to reduce electric rates. What will be the effect of lower electric rates?

Will the metropolitan area be able to capture a greater share of hydroelectric power from the New York Power Authority? This is related to the previous issue.

Effects of Implementing the Clean Air Act Amendments. Measures to decrease ozone in the metropolitan area may significantly affect energy supply and use.

Continued ability to meet growth in electricity demand by conservation. Will conservation and demand-side management penetrate the market at projected levels, or will historical nonprice-related hurdles result in lower market penetration with an associated need for more electrical generation?

Will natural gas or district cooling be accepted as substitutes for electric air conditioning? These are important options for decreasing overall electricity demands and especially critical peak demands, and decreasing greenhouse gas emissions.

Will the Federal Government impose a requirement that the states take action to reduce greenhouse gas emissions as part of the Framework Convention on Climate Change?

To what extent might any future restrictions on carbon emissions from the metropolitan area be met by joint implementation with other parts of the world? For example, to what extent can nonenergy measures such as reforestation offset energy emissions or satisfy future requirements to reduce energy emissions?

Will current public attitudes that prevent introduction of more nuclear power continue? Or will it be possible to use nuclear power as a means to reduce fossil fuel use?

Will New York State be able to import hydroelectric (or nuclear) power from Canada in the future when (and if) it is needed?

What will be the effect of further structural and institutional changes in the electric utility industry?

Will (or when will) an economic low-carbon-emission automobile become available?

NEW YORK'S ABILITY TO REDUCE GREENHOUSE GAS EMISSIONS

New York's ability to reduce greenhouse gas emissions was estimated using the two-region New York State model with results shown in FIGURES 1 and 2. The upper curve in FIGURE 1 indicates the growth in emissions of carbon dioxide—the principal greenhouse gas—in the absence of constraints. The model was also constrained to limit carbon dioxide emissions to the 1988 level, as indicated by the level line extending from 1998. The most economical sources of the reductions in carbon dioxide are shown in four categories. In 2013, for example, carbon dioxide reductions are due about equally to, on the one hand, energy conservation and energy efficiency, and, on the other hand, fossil fuel shifts away from coal and oil and toward natural gas. Subsequently, there are contributions from additional renewables and, beginning in 2013, nuclear power.

Potential contributions from energy conservation measures are generally modeled in three successive stages at increasing cost. In most cases, the first stage is cost-effective even in the absence of carbon dioxide emission restrictions, and therefore does not appear in the figure. For residential and commercial buildings, the potential energy savings from all stages of conservation is assumed to increase linearly over time, reaching a maximum of the order of a 50 percent reduction by 2028 (ref. 9, Tables A3, A21, A31, A49, A53).

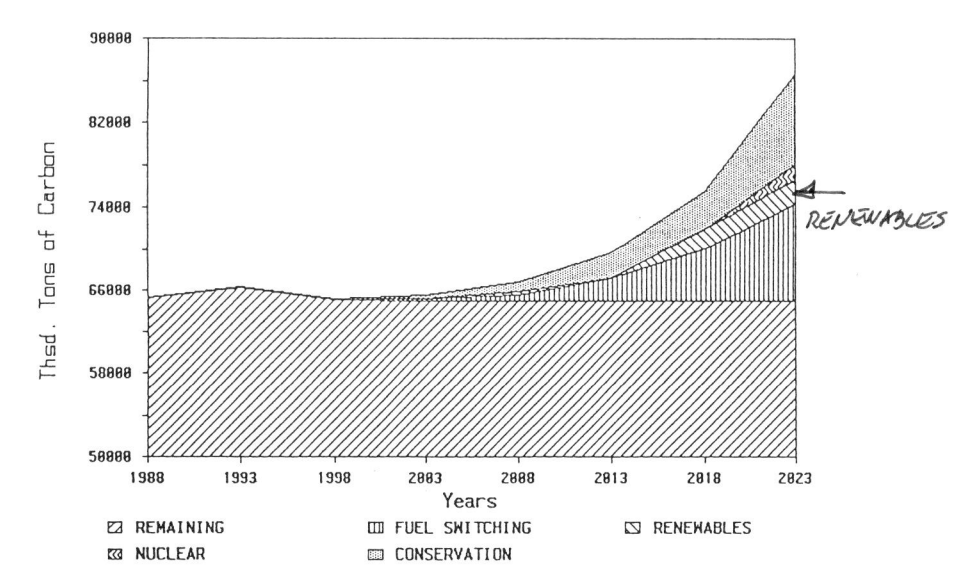

FIGURE 1. Sources of New York State reductions to stabilize CO_2 emissions at the 1988 level.

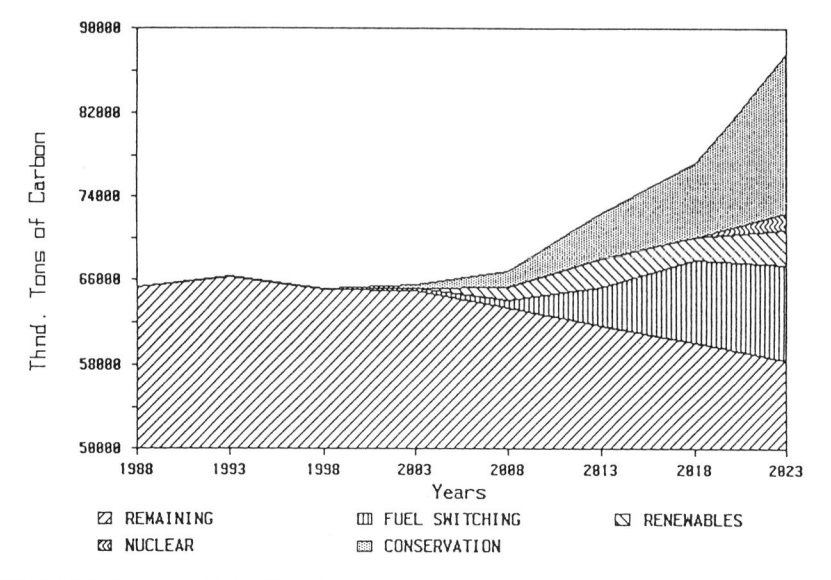

FIGURE 2. Sources of New York State reductions to reduce CO_2 emissions by 10 percent from the 1988 level by 2023.

In a second step, carbon dioxide emissions in the model were assumed to be reduced from 1998 to 10 percent below the 1988 level in the year 2023 (FIG. 2), albeit at a higher cost. With this restriction, the reduction in emissions due to conservation about doubles, producing most of the additional savings. Reductions from fossil fuel switching increase by about half in the 2013–2018 time period, requiring more natural gas than current commitments. There is little increase in nuclear energy, but renewables become an important source of emission reductions.

New York's Ability to Change Energy Demand to Reduce Greenhouse Gas Emissions

New York State was cited as having the lowest energy intensity (MBtu/GSP) among the 50 states (NYSEO, 1994b, p. 17). New York City itself enjoys high energy efficiency. Multifamily housing is more efficient than single family. There is an excellent mass transit system. Pedestrian traffic, the most energy efficient form of transportation, is high. In Manhattan there is a district heating system, and many large building complexes throughout the city and the metropolitan area use cogeneration. The suburbs are not as energy efficient, but progress is being made. Building codes require more efficient and tighter new housing. High energy prices have driven owners to make improvements in energy efficiency of existing buildings. Appliance efficiency standards and demand-side management (DSM) programs are improving efficiency of electricity use. New York's INFORM system and high-occupancy vehicle (HOV) lanes are improving efficiency on the highways. Much improvement is still needed on mass transit in the suburbs.

New York utilities expect that electricity savings associated with demand-side management activities will increase at an annual rate of 10%/year from 1993 to 2008 (NYSEO, 1994b, p. 26). Many opportunities for energy conservation still exist. Examples of energy conservation opportunities are given in TABLE 3.

New York's Ability to Change Energy Supply to Reduce Greenhouse Gas Emissions

Much of the growth in renewable energy in New York State would be enhancements to existing hydroelectric or new, low-head hydroelectric. Other renewable technologies included in the analysis were wind, solar thermal, photovoltaics, biomass, biogas and municipal solid waste. Limits were placed on these technologies to constrain their potential to what were considered realistic levels. Under carbon emission constraints, all of these technologies reached these upper bounds. Model results indicated that wind, photovoltaics, and solar steam electric were the most consistently attractive, although opportunities were more limited in New York than in much of the rest of the U.S.

In 1992, renewables represented 9.2% of primary energy consumption in New York State. It is estimated that the achievable potential of additional renewable energy as a source of electricity in New York State is 1,000 to 1,850 megawatts (NYSEO, 1994b, pp. 70–72). The higher value is 5.4% of the 34,212 megawatts of total installed electrical generation capacity in New York State in 1992. A breakdown of potential carbon reduction is given in TABLE 4. In considering the maximum reduction from the options in TABLES 1–3, note that 1987 carbon emissions in New York State were 63 million tons.[5]

Opportunities for Carbon Sequestering

An alternative to reducing carbon emissions is to sequester carbon. There are three basic approaches: 1) capture carbon dioxide from the air by growing biomass

TABLE 3. Demand-side Opportunities for Reducing CO_2 Emissions

Carbon Reduction Measure	Maximum Reduction (million tons C/yr)	Cost ($/ton C reduced)
Ethanol use in motor vehicles	0.84	105.
Industrial heat pumps	0.4	136.
Adjustable speed drives/high efficiency motors	0.82	5.
Electric vehicles	0.2	418.
CNG vehicles	0.2	10.
Traffic control measures	0.4	15.
CAFE standards to 42 mpg in 2000	3.3	2.
Mass Transit	2.4	500.
Heavy duty truck CAFE standards	1.2	3.
District heating and cooling	1.2	20.

SOURCE: New York State Energy Office, 1994c.[4]

TABLE 4. Opportunities to Reduce Carbon Emissions in Energy Supply in New York State

Carbon Reduction Measure	Maximum Reduction (million tons C equivalent/year)	Cost ($/ton C reduced)
Replace coal-fired generation with fuel cells	0.9	214.
Wind power	1.11	245.
Landfill methane recovery	0.6	31.
Biomass electric	0.43	136.

SOURCE: New York State Energy Office, 1994c.[4]

and then sequestering the biomass; 2) capture carbon dioxide emissions from fossil fuel combustion and sequester them; and 3) remove all or part of the carbon from fossil fuels, sequester it, and burn the remaining low-carbon or carbon-free gas (hydrogen). The cheapest and simplest is the first. It is usually done by planting trees (or by preserving trees that would otherwise be cut down). Carbon is sequestered for the life of the tree, and, assuming the tree is converted to building material or furniture, a substantial portion is sequestered for a considerably longer time.

A New York State study estimated the maximum opportunity for carbon sequestering through reforestation as shown in TABLE 5.

AIR CONDITIONING AND CLIMATE CHANGE

Impacts of climate change are likely to result more from changes in extreme weather conditions than from an average increase in temperature. Global warming in the metropolitan New York area is apt to result in more extremely hot days (over 95°F or 35°C). Epidemiological studies have shown that such days cause increased mortality, primarily among the elderly. Under conditions of global warming, there will therefore be public health reasons to increase the number of residences with air conditioning. In addition, many people in the metropolitan area who do not have air conditioning would install it for comfort. The increased air

TABLE 5. Carbon Sequestering by Reforestation: Potential and Cost for New York State

Alternative	Land Area (1,000 acres)	Total Carbon Sequestered (million tons/yr)	Cost ($/ton C)
Tropical reforestation	500.	0.86	6.
Plant on dry pasture land	53.	0.12	20.
Plant on dry cropland	172.	0.43	26.
Plant on wet pasture land	192.	0.37	27.
Plant on wet cropland	78.	0.17	38.
Public forest management	525.	0.13	150.
Private forest management	1,543.	0.39	200.

SOURCE: New York State Energy Office, 1994c.[4]

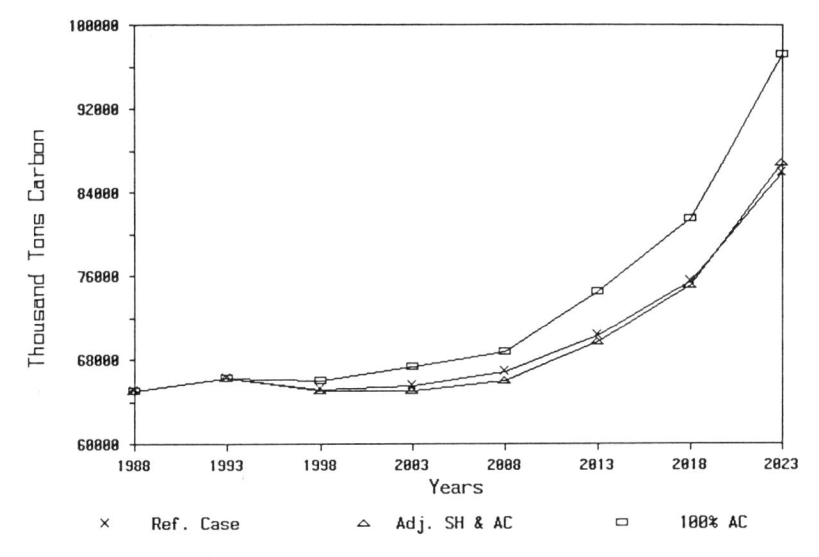

FIGURE 3. Effect of changes in air conditioning on New York State CO_2 emissions.

conditioning load would have two effects: First, it would add to the absolute amount of heat, since indoor heat is pumped outside with additional energy loss turned into heat. Second, it would increase the demand for electricity and the need for electrical capacity (since the Metropolitan area has a summer peak). This would lead directly to increased carbon dioxide emissions, since the bulk of the electric power supply for at least the the next quarter century will continue to be fossil fuels.

To examine the potential impact of an increased downstate air conditioning load on carbon dioxide emissions, the New York MARKAL model was run for three cases: the reference case, a case where heating and air conditioning loads were adjusted for expected changes in local temperatures (but not including any increase in air conditioning over the reference case), and a case with air conditioning in every downstate residence (FIG. 3). Adjusting for the temperature change generally leads to a slight decrease in carbon dioxide emissions because the reduction in heating is greater than the increase in cooling. With air conditioning in every downstate residence, there would be up to a 10 percent increase in annual Statewide carbon dioxide emissions.

OPPORTUNITIES FOR JOINT IMPLEMENTATION

Studies around the world exploring strategies for greenhouse gas reduction almost invariably find the cheapest opportunity to be replacement of coal combustion with a combination of a lower carbon content fossil fuel (natural gas), with nonfossil energy sources, or with conservation. New York State, along with New

Jersey and Connecticut, do not rely heavily on coal, and coal use in the metropolitan area is minimal. The most cost-effective approach to reducing greenhouse gas emissions is thus not available. Although there are many opportunities remaining for energy conservation, as discussed, New York is already relatively energy efficient compared to other states. It is not surprising, then, that a recent study of New York State found that the cost of achieving carbon emissions reduction targets of given percentages were higher in New York than in the rest of the U.S.[1]

This makes New York State and the metropolitan area candidates for joint implementation of greenhouse gas emissions. This term has been applied primarily to nations, rather than states or cities, but applies equally. It means that an area facing high costs for emission reduction teams with an area facing low costs, and pool their resources to achieve a desired percentage reduction of the combined emissions. It has the same effect as creating tradable rights in greenhouse gas emissions.

COLLATERAL REDUCTION OF OTHER POLLUTANTS

It was noted above that actions under the Federal Clean Air Act to reduce emissions of sulfur and nitrogen oxides, particles, and hydrocarbons also generally act, in many but not all cases, to reduce carbon dioxide emissions. The reverse is true to an even greater extent. The MARKAL analysis suggested that stabilization of carbon dioxide emissions would result automatically in a reduction of emissions of nitrogen oxides beyond the reference case. Carbon dioxide constraints tend to move the system in the direction of overall improvements in environmental quality. On the other hand, it must be remembered that further reductions in conventional air pollution have not been made largely due to the high costs involved. The fact that they may occur automatically as a result of meeting carbon dioxide reduction targets will not make these costs any less burdensome to society.

CONCLUSIONS AND RECOMMENDATIONS

- The most effective actions to reduce carbon dioxide emissions are energy conservation and efficiency improvements.
- To achieve stabilization of carbon emissions over decades, however, these will not be enough. Supply-side changes such as switching to natural gas or renewables will be necessary.
- To achieve substantial reductions beyond stabilization, greater implementation of cogeneration and high-efficiency new technologies such as fuel cells will be needed.
- To achieve very substantial reductions, *e.g.*, 20% or more, considerable addition of nonfossil resources will be needed, coupled with major reductions in the use of fossil fuels for motor vehicles. The need for nonfossil resources can probably be met only by substantial increases in imported hydroelectric or by nuclear energy.

REFERENCES

1. MORRIS, S. C., J. LEE, G. GOLDSTEIN & D. HILL. 1992. Evaluation of carbon dioxide emission control strategies in New York State. BNL 47110. Upton, NY: Brookhaven National Laboratory.
2. NEW YORK STATE ENERGY OFFICE. 1994a. New York State Energy Plan (Draft), Vol. II: Issue Reports. New York State Energy Office, Albany, NY.
3. NEW YORK STATE ENERGY OFFICE. 1994b. New York State Energy Plan (Draft), Vol. III: Supply Assessments. New York State Energy Office, Albany, NY.
4. NEW YORK STATE ENERGY OFFICE. 1994c. Impact Assessment Report: Technical Appendices. New York State Energy Office, Albany, NY.
5. NEW YORK STATE ENERGY OFFICE. 1991. Analysis of carbon reduction in New York State, New York State Energy Office, Albany, NY.
6. CALIFORNIA ENERGY COMMISSION. 1989. The Impacts of Global Warming on California, Interim Report. California Energy Commission, Sacramento, CA.
7. BROCCOLI, A. J. 1996. The greenhouse effect: The science base. Ann. N. Y. Acad. Sci. **790:** 19–27. This volume.
8. NEW YORK POWER POOL. 1993. Load and Capacity Data. Vol. 2. Altamont, NY.
9. AMERICA'S ENERGY CHOICES. 1992. The Union of Concerned Scientists. Cambridge, MA.
10. NEW YORK POWER AUTHORITY. 1994. 1993 Annual Report. New York, NY.

ADDITIONAL COMMENTS BY AJAY SANGHI

I am the project manager of the New York State Greenhouse Study which is just being launched. The study is partially funded by the U.S. Environmental Protection Agency. We consider that it will ultimately become an integral part of the energy planning process. The energy planning process in New York State is unique. Even though New York State Energy Office is the lead agency, several other New York State agencies work on the plan, and they will also work on this greenhouse gas study. Our time frame is short. We have to finish it in a year, and wrap up all the reports including the public dissemination booklets within 18 months; that's what we have promised the EPA.

We will have a lot of the same kind of thing that Sam did on the technical aspects. And I heard from the speakers price, price, price. I am an economist, and I understand price, price, price. But if you sit in Albany, it's no, no, no. And take it home with you.

Public policy is not governed by the excellence of the science or the economics necessarily. All these measures that Sam and my other colleagues here have shown are good, and we will do that exercise. But is that where we will finish? I don't think so.

The issue is: how can we take a step in trying to do something when you know that you cannot legislate a tax? We in New York State were the forerunners of the environmental externalities concept. And I did all of that work. But understanding the reality and the complexity of our society, we must move ahead. We would like your suggestions, mostly in the area of what different groups in our society can do voluntarily. What educational efforts can we make? How can New York State, the State Energy Office, and the energy plan to be helpful in that? The Economic Development Agency and the Department of Environmental Conservation are partners with us. How can we help to do what can be done, besides saying, OK, if we have this much tax, that can done and that can be done? I will really appreciate your input. If you want to see something done in New York State, this is the vehicle, and don't sit back.

Sam did ask me about the carbon situation. We funded a study by Syracuse University which is reported in our energy plan. We believe that carbon emissions are going to go up by about 2 million tons between 1990 and 2010, and that carbon sequestration potential through reforestation in New York State is about of the same magnitude. You could probably save by planting trees. The costs range anywhere from $6 a ton to about $38 a ton, at an average of about $25. Thank you.

Emission Reduction in New York State

EDWARD O. SULLIVAN[a]

Deputy Commissioner for Environmental Quality
New York State Department of Environmental Conservation
50 Wolf Road
Albany, New York 12233

Lieutenant Governor Lundine asked me to extend his apologies for not attending. He was very interested and serious about attending here today. He has been very supportive of all the activities of the Department of Environmental Conservation, but himself has championed the magnetic levitation initiative that is consistent with the goals of addressing global warming in New York State and beyond.

I would like to commend the American Society of Civil Engineers, Metropolitan Section, for tackling this rather cosmic and important issue, for getting involved, for identifying what you can do in the metropolitan area and beyond, to slow down our movement toward a global warming trend and also to react to the implications of it once they are here with us.

Because the issue is so broad, I am going to focus on energy efficiency in transportation in my remarks. But I think the topic of the greenhouse effect and global warming causes us to focus on the troubling question about how our society, our government, our industry, we as citizens react to situations characterized by a degree of scientific uncertainty. I am reminded of the tanker-truck on the highway barreling along at 65 miles an hour with a warning sign on the side saying, "The scientific community is divided on the toxicity of the chemicals being carried by this truck."

While you may never have seen such a sign on the highway, it is probably true of a lot of the substances that are transported through our communities. I can assure you that if you reflected on it for a moment, and notwithstanding the scientific uncertainty, any of us who found ourselves trailing such a vehicle would hope that the driver is competent, that the truck's brakes are working, that if there were an accident, the "hazmat" team would arrive quickly and know what to do to contain the spill effectively and protect us from the public health and environmental implications of such a spill.

So I think it behooves us to take action to assess what we do know about a situation, and to take deliberate action based on a good evaluation of the scientific information that is available to us. There have been too many situations—I think of the Reagan Administration's go-slow policy on this issue as well as the stratospheric ozone depletion issue—where there seemed to be an inclination to capitalize on the uncertainty. This was reflected in the reluctant posture we took to Rio, where issues such as global warming, ozone depletion, and loss of species diversity were downplayed. Too bad an effort was not made to go to the heart of what we do know, and to forge strategies based on areas of consensus within the scientific community.

---

[a] *Present address:* Commissioner, Maine Department of Environmental Protection, State House Station No. 17, Augusta, ME 04333.

This characterizes the way that New York State has approached this issue. Governor Cuomo really provided a remarkable degree of leadership here. In the spring of 1989, Governor Cuomo, along with then-Governor Kean of New Jersey, and Madeleine Kunin of Vermont, sponsored a four-day conference on global climate change here in New York City. At that conference, there were over 400 scientists and public policy officials from around the world. I had the opportunity to attend that session myself.

The conference was a real working session that went on for a number of days, where people really rolled up their sleeves and reckoned with "what do we know?" and "what can we do?" based on that knowledge. It was aimed at providing a scientific basis for developing a policy position in New York State that we could then extend to our participation in the national and the international debate. Interestingly, back in 1989, there was real consensus among the participants that greenhouse gases introduced into the earth's atmosphere by human activity will force a change in the earth's energy balance causing temperature to rise over the next several decades. There was agreement that there was urgency about addressing the problem and that we need to begin implementing preventive measures and establish a clear state and national energy policy and to continue our efforts in working with other countries to address the problem in all its complexity.

Going beyond and rising out of that conference, New York State has participated as an informed and credible player in the development of policies aimed at increasing our energy efficiency, reducing our energy consumption, and moving toward a more enlightened transportation policy. Among the specific recommendations coming out of the round table are the following: i) We should be working towards a 20 percent reduction in carbon dioxide emissions by the year 2000. ii) We should strive for the elimination of chlorofluorocarbons by the year 2000. iii) We advocate the improvement of fuel economy of new cars to at least 42 miles per gallon, also by the year 2000.

This effort helped to precipitate further actions on a state and national level. A notable example was the United Nations Conference on Environment and Development held in Rio in June 1992 which adopted a framework on global climate change, an agreement to reduce greenhouse gas emissions.

This past April President Clinton announced the nation's commitment to reduce emissions of greenhouse gases in the United States to the 1990 levels by the year 2000. Also, the President committed to produce a cost-effective plan to continue the trend of reduced emissions thereafter. On October 19, 1993 the President made this announcement and released a Climate Change Action Plan which commits the Federal government, in partnership with states and the private sector, to a broad range of measures aimed at stabilizing greenhouse gas emissions by the year 2000. The action plan calls for over fifty initiatives to achieve this goal. Now a detailed plan is being prepared by the administration so that the United States can go to an international framework for this, because in March of 1995, there is going to be an international convention in Berlin where this will be addressed on a global basis.

New York State and its industries are already actively participating in this effort. President Clinton has established an advisory committee that is developing specific action steps to reduce greenhouse gas emissions from cars. This 30-member committee is charged with developing measures that will significantly reduce emissions from personal motor vehicles. Frank Murray, our commissioner of the State Energy Office, and David Freeman, our chairman of the New York State Power Authority, are the Governor's delegates to this effort. The committee will develop recommendations on policies that would most cost-effectively return to the 1990 levels of greenhouse gases from cars by the years 2005, 2015, and 2025. There is a commitment already in place that there would be no increases after that date.

The personal automobile is a critical aspect of this strategy. New York State is participating officially in the context of compliance with the Clean Air Act. This is also an area where each of us can take action through our selection of vehicles and through our transportation patterns.

New York State is also, under the Governor's direction, conducting a study which is being carried out by the State Energy Office to provide an in-depth evaluation of New York's greenhouse gas emission inventory. So we are creating an inventory of all the sources of greenhouse gases in the state, whether building, industry, transportation activities, or utility activities. Then, we are going to look at developing strategies, including energy efficiency, demand-side management, transportation measures, fuel diversity, and tax policy that can be harnessed to bring down that inventory in concert with what is going on at the national level.

One critical area where we are convinced we can make a difference in this area is technology development. We face choices in the development of our strategies. Some of our choices will involve changes in our life style. Others are changes in our technology, advancements in our technology, particularly in the areas of transportation and energy generation. Carbon dioxide is formed as a result of the combustion of fossil fuels, and so we need to seek new technologies that produce power in the most efficient way, which provide alternatives to the use of fossil fuel both for our industrial practices and in our transportation activities.

There are a number of things that Governor Cuomo has proposed and that we at the Department of Environmental Conservation, the State Energy Office, and the New York State Energy Research and Development Authority can do to promote this. Recently, I participated in an initiative to launch an environmental business association in New York State whose goal is promote the interest of the "green" industry. Governor Cuomo has recently proposed the creation of a center for advanced technology for the environment. In fact, the concept for a center for advanced technology is something that is well established already in New York State. There are about a dozen around the state. These are consortiums of the state's research institutes, state agencies, foundations, and industry, aimed at creating jobs in a particular sector, and achieving some other goals like environmental improvement in this case, by creating synergy among government, private, and not-for-profit players.

The creation of an environmental business association and the Governor's commitment to this concept of an advanced technology center, in conjunction with a related proposal that the Governor has made to create a new investment fund for new businesses in the environmental field—particularly, small businesses—has the potential to create a real wave of investment and development in new technology that can help us address the problem of global climate change.

Transportation plays a critical role in this issue because it is the source of 30 percent of New York's as well as the nation's carbon dioxide emissions. On the transportation front, New York has taken a number of steps which I think are going to be critical in this area. First, as part of our efforts to meet the Clean Air Act Amendment of 1990, New York has adopted the low-emission vehicle requirements that were enacted in the state of California. California has traditionally been a leader in the control of emissions, primarily because they have had the worst air quality problems in the country. In response, Californians have put themselves on the cutting edge of low-emission vehicle requirements, and the Clean Air Act has provisions that allow other states to opt into adoption of

California's standards. Those standards gradually and progressively ratchet down auto emissions, and ultimately require the development of zero-emission or electric vehicles. There is an initial sales mandate of two percent sale of zero-emission vehicles in the year 1998. That increases progressively to ten percent by the year 2003.

New York has been on the front line of this effort. We have been working with all of the states on the eastern seaboard who are part of the Ozone Transport Region. This organization, created by the Clean Air Act, recognizes that bad air moves from one state to another; it doesn't recognize state boundaries. As a result, states within the region have to work together to develop strategies. For example, we are working together to encourage EPA to require the sale of low- and zero-emission vehicles pursuant to the California program in the entire Northeast. We are fighting daily—the Commissioner today is down in Washington—trying to persuade EPA to approve a petition we have before them to impose a low- and zero-emission vehicle requirements on the northeastern states, and to deflect the efforts of the auto industry in this country to curtail this effort. I think that the auto industry is going to be left in the dust, so to speak, if it doesn't realize that low-emission vehicles and zero-emission vehicles are the future for this country. The problem of global climate change brings it home as an issue that they should be concerned about.

We are seeing some progressive developments within New York State. Niagara-Mohawk, a utility in the Syracuse area, for example, has been a real champion of progressive policy in this area. In addition to sponsoring conferences themselves on how to take the lessons of the agreements that were reached at Rio and incorporate them into state and industrial policy in the state, they have made history by entering into a joint venture with a California-based company, U.S. Electric Car. That agreement will provide for construction of an electric vehicle manufacturing facility in central New York that will produce a thousand cars per year. Niagara-Mohawk has agreed to serve as their agent in New York State, to help them market their vehicles, particularly to large fleets. And then the New York Power Authority has committed to spend a million dollars a year to purchase the output from that facility. Our Department of Economic Development is intimately involved in helping U.S. Electric Car make the decision and get them together with Niagara-Mohawk.

This is the kind of synergy that we can realize through a coherent environmental, industrial, and economic development policy in New York State that will help us combat the greenhouse gas and global warming problem. Similarly, Lieutenant Governor Lundine has shown real vision in personally championing the concept of "maglev" rail system in New York State. Magnetic levitation represents a major advancement in rail transportation that is critical to pushing us to a new level of use of mass transit. New York will be the first state in this country to actively develop, build and utilize a maglev system as part of its transportation network. This will represent a real advancement. The Lieutenant Governor has personally been down in Washington getting support for development of a pilot project that is going to be based at the Stewart Air Base in Newburgh.

Energy production is the other area where we have to really focus our attention, because in producing energy we create significant greenhouse gases. Obviously, we want to have sufficient energy for development in the U.S. and for development in the less developed areas. But it is critical that our energy production be as efficient as possible and that we move toward reductions in our carbon dioxide emissions in this sector as well. The State Energy Plan, which was just finalized this past Monday, directly links energy demand and degradation of our air quality and makes a serious commitment to reduction of air quality impacts from energy consumption. The plan calls for implementation of cost-effective demand-side management programs and supports strengthened appliance energy standards and building conservation construction code programs on the state and federal level. The plan's goal is to achieve statewide electricity reduction of 8 to 10 percent from projected levels by the year 2000 and 20 percent by the year 2012.

Energy efficiency remains the cornerstone of New York State energy policy because it reduces cost, improves air quality and bolsters the economic competitiveness of business and industry. Utility demand-side management and state energy efficiency programs have already reduced energy consumption in New York by 6.6 percent since 1992. So we think that an 8 to 10 percent goal remains achievable. But we are going to have to fine-tune our DSM strategies to make sure that there is good collaboration between government, utilities, and industry and the manufacturing sector so that energy efficiency and appliances and equipment and building codes can be part of the strategy.

In closing, I just want to commend this organization for the important role you are playing in two areas. I recommend that you focus on strategies that in the metropolitan area can push demand-side management, building code modification, and use of alternative fuels in the city fleets and in commercial fleets. The city has made a real commitment to the use of compressed natural gas in its fleet. While there was a recent *New York Times* article indicating that the process was taking longer than expected, I think we need to be patient and supportive. We must look for synergy, for opportunities for linkage between the city and pioneering utilities like Brooklyn Union Gas. Once we have our own house in order within the state, we can move to tackle the international issues as well.

The final area where you can make a difference is in education. Unless people have an understanding of the problem, there isn't going to be the willingness to support our state and city programs, or additional scientific research.

DISCUSSION OF THE PAPER

ROGER HERTZ (*Bicycle Transportation Action*): Two questions. I am curious to know how many of the people in this room use bicycles for transportation. And how many people walk for transportation? I am curious, particularly in the light of the Lieutenant Governor's strong support for the good job that the state has done in the last year and a half on bicycle and pedestrian transportation that there was no reference to that in your remarks. I hope in the future you'll consider putting it in. The program was a joke two years ago. It is now one of the twelve best in the country.

EDWARD SULLIVAN: I agree. Thank you very much for noting that omission. Actually, it is a program that I personally have participated in. I am a big biker. I live 25 miles from my job, but I do bike to work sometimes in the summertime. I have been working very closely with the Department of Transportation to advance the development of bike paths as a part of our greenway development process in the state. The Department of Transportation is actually doing a good job of being a catalyst in this area. I have personally been involved with working with the Albany Service Corps which is one of a number of service corps around the state that are making a commitment to development of bike paths and related facilities to help make bicycle transportation and pedestrian transportation a real viable option for people. Thank you for noting that.

MARTIN GARRELL: Given the fact that no matter what New York does, and no matter what California does, the global warming problem is, after all, global. And the fact that from all the presentations in this conference and from modelers' predictions, we are expecting some problems for the next century, do you see any support from the state—financial or otherwise—for things that are other than the proactive things? In other words, so you see support for some of the measures we were talking about this morning along the shorelines? And the monetary expenditures that would be necessary, is there some awareness of that at the state levels, too?

EDWARD SULLIVAN: I think, frankly, our vision has been focused mostly on preventive measures. But I think the day will come when we will have to start thinking proactively about how to protect the shoreline and so forth. That's something that we should take on in conjunction with the Corps of Engineers. The Department of State will obviously play a critical role here. I think it is something that we should begin planning for right now. I am not aware of any active expenditures right now, but I think it is something that we can certainly begin to plan for.

Remarks by Manhattan Borough President

RUTH W. MESSINGER
*President of the Borough of Manhattan
Executive Division
Municipal Building, 19th Floor South
New York, New York 10007*

First, I want to thank and acknowledge the sponsors of this conference: the New York Metropolitan Section of the American Society of Civil Engineers; the Marine Sciences Research Center at S.U.N.Y. Stony Brook; the Institute for Marine and Coastal Sciences at Rutgers University; the New York Academy of Sciences; the Regional Plan Association; the Port Authority; the Sea Grant Institute; and the New York State Bar Association's Environmental Law Section. I also want to acknowledge the work of Douglas Hill, who spent valuable time organizing the event and working with my staff. Finally, I want to thank my long-time friend Philip Jessup of the International Council for Local Environmental Initiatives, who launched the International Cities for Climate Protection Campaign two years ago at the Municipal Leaders' Summit on Climate Change and the Urban Environment at the United Nations.

This is an impressive gathering on an important subject. Your work today promises to be extremely useful to those of us responsible for making and enforcing environmental and land use policy. I look forward to seeing the results of the scenario planning workshops especially, since that promises to be groundbreaking work on how to bring greenhouse solutions into operation in the city and region. I would be interested in aiding in taking further action on these results through the forum of the National League of Cities.

I will not presume to address engineers on engineering questions, or lecture environmental scientists on environmental science. The first rule in public speaking, as it is in writing, is: stick to what you know.

So I will stick to what I know: public policy. Specifically, I will touch on two subjects: the range of policy options pertinent to the greenhouse effect that are available to local government; and the mechanisms that would encourage the adoption of such policies.

Despite uncertainties about the magnitude of the consequences of the greenhouse effect, there are no mysteries about what is producing the it, and no surprises in the public policies required to contain it and contain its consequences.

We all know what those policies must achieve: increased energy conservation; water conservation and watershed protection; encouraging mass transit and other less and non-polluting transportation systems; waste reduction, recycling and reuse; land use and development policies that preserve our shorelines and set energy; and resource-efficient design and construction standards. Taken together, they are the building block concepts of sustainable development, a concept my office applies to its work.

One example of that work is in the area of waterfront planning, and it relates directly to issues raised at this conference.

In a few weeks, I will send to the City Planning Commission a proposed plan for the entire thirty-two mile waterfront of Manhattan. Under the New York City Charter, the planning commission must review and act on that proposed plan; if the plan is adopted, it becomes city policy. The plan demonstrates that the Manhattan waterfront is a great untapped resource—or, to use another metaphor, a neglected but still valuable heirloom—for New York.

How does this relate to the greenhouse effect? If adopted, our plan would encourage alternative and mass transit systems that would reduce carbon dioxide emissions from trucks and autos. The importance of encouraging such alternatives cannot be exaggerated. Each day, more than three million people enter and leave Manhattan's central business district—far too many of them in the automobiles that make Manhattan an air pollution hot spot. Tim Forker, an environmental policy specialist on my staff, who has been part of this conference, informs me that New York City is not only the number one source of carbon dioxide emissions in North America—a dubious enough distinction—but that carbon dioxide emissions here exceed those of Mexico City, Los Angeles, Chicago, Dallas-Fort Worth and Houston—combined.

The plan addresses the issue of reducing carbon dioxide emissions several ways. Much of the plan focuses on the waterfront's potential for a continuous esplanade and bikeway around the island, which we hope would become an alternative commuting path for many New Yorkers. In addition, we urge far greater use of ferries and waterborne cargo transportation in and out of Manhattan, an area in which New York City lags behind other cities.

Today, only about eighty thousand commuters use ferries crossing the Hudson and East rivers. But ferry ridership has grown steadily in recent years. Last month, new high-speed ferry service from Hunters Point Queens to East 34th Street was inaugurated. In the next few months, similar ferry routes will begin operating from Rockland County and Staten Island to West 39th Street.

Government must encourage development and licensing of such services. We must find the dollars to create and improve ferry landings, as has already been done with federal "ISTEA" monies. We must coordinate ferry service with existing or planned mass transit systems both in and outside Manhattan. Bus lines that now terminate blocks from the water's edge, for example, need to be brought out to the shoreline to link up with ferry services. The same holds true for the projected 42nd Street "trolley." Maps, advertising, promotional campaigns need to be redesigned to promote ferry travel.

As exciting and innovative as such new transportation services are, it's crucial to maintain investment in the existing mass transit system, which serves some five million riders a day. Unfortunately, recent cuts in the MTA capital budget threaten to undo the hard-won progress we've made in relieving the worst flaws of the transit system. And in the operating budget, funding cuts undermine the potential of the Metrocard and other innovative pricing strategies to increase ridership and tie the elements of the transit system more closely together by offering off-peak discounts or by reducing or eliminating two-fare zones.

As to my second major point: encouraging the adoption of policies that mitigate or contain the greenhouse effects.

As this conference has made abundantly clear, the greenhouse effect does not respect political or jurisdictional boundaries. Its impact will be regional—on farmland, on rivers and wetlands, on upcountry watersheds, on coastal communities, on city dwellers.

Government resources are limited, and government's powers are further limited by the difficulties of mobilizing action across political jurisdictions. And even if regional procedural or institutional mechanisms existed, they would not themselves be sufficient absent a political commitment to making them work. Time and again, we have seen failures to develop regional economic development strategies, for example, because of the absence of such political will.

How then to build political support for regional action? Look to your left; now look to your right. You've just met the people who have to help make it happen. One place to begin—an idea I'm sure that Phil Jessup endorses—would be for New York City to join the International Cities for Climate Protection Campaign, which he launched nearly two years ago at the Municipal Leaders' Summit on Climate Change and the Urban Environment at the United Nations. That could be a recommendation of this conference.

Another suggestion is that you begin developing a system of generally accepted indicators, or "trip wires" that would suggest the need for well-calibrated and appropriate responses to changing environmental conditions. When mean sea levels reach certain heights, or seasonal degree days rise to particular levels, or saltwater intrusion of the Hudson and Delaware rivers passes certain points—these can be indicators that discrete government responses throughout the region are required. Public officials badly need such a system for managing decisions and potential actions.

By the way, my waterfront plan also calls for establishing educational and research centers on the Hudson River piers that would be open to the colleges and universities in the region, and that I hope would be used for such research and monitoring.

Finally, government, business and professional associations can do a great deal more to energy-efficient building design and construction standards—to encourage conservation of resources—to underscore the imperatives of waste reduction and recycling and use of renewable energy sources. However, as you heard yesterday, there are important but unconnected planning efforts underway now across the region to deal with global warming. I'd like to join you in better coordinating those efforts.

Government can and does make a difference, both for good and for ill. From my offices at the Municipal Building, I have a view of Southbridge Towers, middle-income housing that has the largest solar heating system in the city. It was completed with a grant from the Federal Housing and Urban Development Agency.

That was in 1981. For ten years after that, the federal government effectively ignored development of solar and other non-fossil fuel energy systems. We raised the speed limit and reveled in an oil glut—and pushed the global thermometer steadily upward.

This conference demonstrates that we are ready, as a society, to resume a more sober and reasonable path—one that recognizes our responsibilities to each other and to future generations.

In the current issue of *Wilderness Magazine*, Bill Yellowtail, of the Crow Nation, writes: "Ecosystem protection. Sustainability. Environmental democracy. Right. Finally, you're getting it. But American Indians are not much for saying 'we told you so.' We're just glad because maybe it's not too late, yet." For all our sakes, let's prove him right.

DISCUSSION OF THE PAPER

JEANNE FOX: Is that the Bill Yellowtail who is a regional administrator for the EPA?

RUTH MESSINGER: Yes.

MALCOLM BOWMAN: Can you tell us a little bit about the city's recycling program? They say it is under threat from budget cuts.

MESSINGER: I was afraid somebody would ask that because I think it is a really good example of the opposite of what I am talking about. The City Council, when I was on the City Council, originally passed a local law requiring recycling and set in fairly high expectations and tonnage levels and made that the law of the City. The whole concept in any area for passing a law is to require changes in behavior. What we hoped we could count on the last three municipal administrations to do was to back up that law with a fairly thoughtful system of resources, outreach, education, and enforcement.

For example, in our office a year or two ago, we said this is ridiculous. You cannot enforce a recycling law in which the instructions come on four printed pages and every community board in the Borough of Manhattan has different instructions for different materials on different days. Because there is no way to get that information out. So that was the bad news. The good news was, they all went to all seven materials. There was a fairly even promotional effort; we received information for maybe three months, six months. Now we are going to start enforcement. We said, this just doesn't make any sense. We have huge areas where people need information in another language, where people need hands-on education. Give us some resources, and we will do education outreach. In the last City administration under David Dinkins, we won that point with Commissioner Emily Lloyd.

We got dollars moved from the Department of Sanitation to our office, a very small sum, $180,000 a year. And Tim Forker supervised six people who went out and did—you can't imagine what they did—they went to tenant meetings. They went to the New York City Housing Authority and told them that it would be a good idea if they complied with the law. And then they educated tenants on how to do it. And they really got people in different kinds of neighborhoods across class and race and neighborhood lines excited about a commitment to recycling. We began to make it work in the only way that we think it can. People won't recycle if they have to walk four blocks in the wrong direction to find a dumpster that's locked. So we tried to coordinate all the agencies and put the pieces together. We were particularly trying to push the City, for the Housing Authority, the Board of Education, and the Health and Hospitals Corporation to recycle, because if they did a perfect job, I promise you tonnage numbers would go way up. If our Municipal Building did a better job, tonnage numbers would go up.

But they undermine their own effort. Even during the year and a half that we had these coordinators, once every four months you read in the newspaper: "Recycling on the chopping block." I mean, the notion that New York City residents are either ostriches with their heads in the sand, or are stupid, is really insulting. And when people read that recycling is on the chopping block, it undoes months of great public service ads on television that say, this is the law, it is good for your health, it will save money, yes, but they might not do it next week, so why should I take it seriously?

So they kept threatening the program, and then they did away with the coordinators, they did away with the outreach, they threatened in June—but did not—and are now threatening again to go—I just find this so appalling, it's hard to talk about—to go to once every two weeks instead of once every week pickup in the lowest complying districts as a way of punishing them. Folks, the lowest complying districts in Manhattan are the lower East Side, west Harlem, east Harlem, and Washington Heights-Inwood. And in addition to the obvious language and education problems that it will take to get those neighborhoods to increase their tonnage, there are huge problems in all of those areas of population density and limited building storage. So I am sorry to tell you that the one thing that we think we can promise the City is that if you double the wait before recyclables get picked up, that fewer people will make an effort to recycle because there is no place to store the material. And frankly, if they are stored, then we will have a renewed and reinvigorated rat population enjoying the recyclables, because we cut back on pest control. Anyway, that's the specific that is under threat right now. I don't know how many districts it is across the city, but it is five out of twelve in Manhattan would go to once every two weeks.

BOWMAN: It is a real shame because right now the market for recyclables is exploding for paper, aluminum cans. Aluminum cans are $800 a ton. You can make money. You can save an immense amount of money.

ROBERT ALPERN: I have been obsessed with the concept of adaptive management. Your notion of trip-wire indicators sounds like something along those lines Ruth, I wonder if you could think a little more aloud about adaptive management, avoiding premature actions, or untimely actions that are too late. How might we move toward the adoption of trip-wire indicators?

MESSINGER: The simple answer to your question, Bob, is I haven't thought about it. I don't think that immediately is a consideration. The reason I talked about trip-wire planning is—and maybe this is speculative, and maybe it is incorrect speculation—but sometimes having people with technical information in agencies, in professional associations, or in lobbying groups, come to people in government and say, this is dire, and you must do X right now. You know what that leads to: an immediate cost-accounting, we can't afford it, so we'll pretend it isn't, we will undo Local Law 19 which is the recycling law, so that we won't be out of compliance with our own law.

I just thought that taking an issue, which is what I hope will come out of the scenario planning efforts, that say on carbon dioxide emissions, or on something else, here is the number where we are right now; here is the number where, if we hit it, we can promise you there would be the following additional negative consequences unless you did X. Here is another threshold. That might allow people a little more time to get ready for it. It would mean that if we stayed at the current level—while there might be many things it would be desirable to do to reduce that number—we would not be incurring an immediate additional cost. If we saw ourselves climbing up, we would have to deal with the fact that, either we are going to keep climbing and we would hit a need for some significant dollar spending months or years down the road. Or we would have to put in new measures that would get us to bend back away. I just saw it as maybe a little bit less dramatic and draconian than saying, you've already incurred such . . .

The watershed is not a bad example (since you and I have walked it). I actually think that is a good example. People do understand that if some set of bad things happens, it's going to cost this city $8 billion. I am not saying that's any longer the number, I am saying, that's what people understand. You push an elected official, you say watershed or filtration, they say $8 billion. They make the connection. They've made that in their minds. And the result is that things that bode well or badly for watershed protection sort of carry that price tag, not as a dollar amount you have to spend right now, but that as something that is looming out there. I think it is what led the last administration to appropriate several hundred million dollars in a tough economy toward purchase of land and for the protection of land. I can't imagine what is in the minds of the people who have moved away from that agreement. But at least as a trip wire, that is a notion that is there. Maybe there are others like that.

DOUGLAS HILL: Well, Ruth, as a group here, we have been spending a lot of time in 2020 in the last 24 hours. Thanks for bringing us back to 1994.

MESSINGER: Please let me know what is going to happen in 2020, but break the news to me gently.

Introduction to Scenario Planning Results

MALCOLM J. BOWMAN

Marine Sciences Research Center
State University of New York at Stony Brook
Stony Brook, New York 11794-5000

Welcome to this last phase of our meeting. My responsibility is to introduce our facilitators and rapporteurs, but before I do that, I will explain to you what scenario planning is, in case you're not sure.

What is scenario planning? Scenario planning is really role playing. We have asked our participants in this two-day workshop to each put on a hat, to pretend that perhaps that they are somebody else, and to assume a set of positions that they might not necessarily ágree with. But by doing this, by role playing, perhaps we will see the problems associated with global warming in a different context.

Our participants are quite diverse. They include academics and their students, state planners, engineers, lawyers, economists, and environmentalists. We have an international representation from the United States, Canada, the United Kingdom, China, and continental Europe. We've been thrown together for these two days. Many of us are what might be called intelligent lay persons. Some of us are experts, some are professionals working in government, industry, universities or in consulting. But the issues surrounding global warming are so broad and so profound that most of us don't claim to be experts.

The three groups you have been randomly assigned to are:
Scenario One. There will be no change in the climate of the earth due to global warming over the next fifty years.
Scenario Two. Moderate change. If you look at TABLE 1, you will find scientific estimates provided by Dr. Broccoli of Princeton University that have been used to predict what the magnitude of global change might be, relevant to Metropolitan New York's infrastructure.
Scenario Three. An accelerated change in global warming through the year 2020. For example, look at the *Medium* column and the row for *Sea Level Rise* in 2030,

TABLE 1. Climate Assumptions for Scenario Planning

		IPCC Estimates			
Expected Climate Change →	No	Low	Medium	High	Accelerated
Global Average Temperature					
2030	+0°F	+1.3°F	+2.0°F	+2.7°F	+3.4°F
2070	+0°F	+2.9°F	+4.3°F	+6.9°F	+8.9°F
Local Temperature					
2030 Winter	+0°F	+2.7°F		+5.4°F	+6.75°F
Summer	+0°F	+1.8°F		+3.6°F	+4.5°F
2070 Winter	+0°F	+5.4°F		+10.8°F	+13.5°F
Summer	+0°F	+3.6°F		+7.2°F	+9°F
Sea Level Rise					
2030	+4.25 in	+7.5 in	+11 in	+15.5 in	+19.5 in
2070	+9 in	+17 in	+28 in	+37 in	+47 in
Precipitation					
2030 Winter		+0%		+15%	22.5%
Summer		−5%		−10%	−12.5%
Summer Soil Moisture					
2030		−15%		−20%	−22.5%
Frequency of showers and thunderstorms			—more—		
Day-to-day and interannual variability of midlatitude storm tracks			—less—		
Tropical storm frequency and intensity			—more—		

SOURCE: "Low" to "High" are from Anthony J. Broccoli (22 July 1994), adapted for the New York metropolitan regions from the "Central North America" region in the 1990 report of the Intergovernmental Panel on Climate Change. Sea level rise due to global climate change is added to existing rate of 3 mm/year.

and read "plus eleven inches." And in the *No* climate change column, you will note a sea level rise of 4.25 inches. In other words, sea level is increasing anyway and is increasing owing to factors other than global warming, for example, continental tilting. For *Accelerated* climate change shown in the right hand column, the sea level rise by the year 2030 is listed as 19.5 inches above present levels. The same holds for precipitation, temperature changes, frequency of storms, and so forth.

This is the scientific base that we take as given. Our job is not to debate which of these three scenarios will actually happen. You are given your assignment, you assume that it is going to happen, and you study the consequences.

We hope that this afternoon's exercise will be an interactive one, that it won't be a series of speeches, but will focus on the presentations by each group leader. None of us is naive enough to think we are going to solve all these problems. What I see coming out of this is an exercise in consciousness raising. We're going to leave here with a new awareness, and maybe we'll change the way we think of Metropolitan New York and the problems that we are going to face in the greenhouse era.

Of course, sensible people hope that global warming is not going to happen, that the accelerated sea level rise will not occur. There would have to be draconian measures taken to take care of those problems.

Before I ask the first group to come forward, I should point out just how we went about designing this exercise. If you look at TABLE 2, you will see parts labeled *Categories, Desired Conditions in 2030* (or *"Goals"*) (TABLE 2a), *Predetermined Elements* (or *"Givens"*) (TABLE 2b), and *Critical Uncertainties* (TABLE 2c). And then down the left-hand side you will see the various categories that we have been studying: *Economic and Population Projections, Waterfront Planning, Land Use and Infrastructure Planning, Water Resources and Wastewater Management, Air Quality and Health Effects, Transportation Planning, Building Design and Maintenance,* and *Energy Demand and Supply*. These summaries were prepared by our speakers on the basis of the material they presented yesterday morning and afternoon.

The *Desired Conditions* are what we assume are the desired conditions for Metropolitan New York in the year 2030. *Predetermined Elements* are things that we have assumed will take place. *Critical Uncertainties* are events about which there is considerable debate and uncertainty, and which need to be discussed. We will use all this information to plan what should be done to prepare for the next century.

[TABLE 2 is presented on page 166 *et seq.*]

TABLE 2

Categories	(a) Desired Conditions in 2020, "Goals"
Economic and Population Projections	• Increased economic well-being (Gross Regional Product per Capita) • Real gain in per capita personal income • More equitable distribution of economic opportunity among region's residents • Competitive advantage in global and national markets for high-value-added service and merchandise exports • Accounting of environmental externalities in market-based pricing (*e.g.*, congestion pricing) • No further significant loss in stock of region's natural resources and quality-of-life assets
Waterfront Planning	• Publicly accessible waterfront • Revitalized waterfront, balancing economic development with natural resources and public access

(b) Predetermined Elements, "Givens"	(c) Critical Uncertainties
• Increased population diversity and multiculturalism • Greater immigrant and minority representation in labor force • Automation displacement of routine producers in white collar sector • Advent of knowledge-based compensation • Social transformation to working less and living longer • Corporate re-engineering of workplace (downscaling) • Evolution of a contingent work force • Cost-driven bifurcation of labor force demand (toward symbolic analysts and in-personal servers, away from routine producers) • Increased telecommuting from home or satellite office • Increased labor force demand/supply mismatch • Further disparity in income distribution	• Continued increases in foreign immigration to the U.S. and regional destinations • English language and functional literacy levels of adult work force • Cost/value competitiveness of the region's economy in global markets • Service sector productivity increases • Extent of virtual corporation/boutique team approach • Increased multinational business representation in the region by direct foreign presence or portfolio investment • Changes in consumption patterns • Material source, throughput, energy and waste reduction in production processes • Capital availability for public infrastructure and private sector investment
• Industrial decline, leaving industrial waterfront properties vacant • Transportation shift: Containerization and the consolidation of shipping operations to containerports, leaving industrial property's piers and waterfrontage unused and often unmaintained	• Land use policy: Will industrially zoned lands along the waterfront be rezoned for commercial, residential or mixed use? • New water-dependent uses: Will industries, such as recycling, emerge as new users of waterborne transportation? Will other new users be identified that could use the water as highways, thereby needing waterfront locations? • Economy: Will there be an economic boom in which the development community will want to redevelop the underutilized lands along the waterfront? • Property owners' desire to be at the shore: Will there be a storm of significant enough magnitude to change property owners' desire to rebuild in a hazardous location?

TABLE 2. (*Continued*)

Categories	(a) Desired Conditions in 2020, "Goals"
Land Use and Infrastructure Planning	• An infrastructure base that can: 　—promote and support a land use pattern that is consistent with global warming mitigation measures for energy utilization 　—support population and economic activity in the region 　—withstand potential storm surges and flooding from weather modification and sea level rise • A process of infrastructure and land-use planning that: 　—explicitly incorporates criteria for energy utilization that is consistent with global warming mitigation measures 　—integrates these measures into the capital programming process, zoning, and land use regulation 　—develops and maintains a data base for infrastructure and land use for the purpose of identifying areas, facilities and activities vulnerable to the effects of global warming

(b) Predetermined Elements, "Givens"	(c) Critical Uncertainties
• The New York region has the highest population density along the shoreline of almost any area of the country. • A very large in-place infrastructure that supports existing land uses in the region, which is vulnerable to flooding and associated destruction at numerous points throughout the region • The use of facilities now in shoreline areas—however vulnerable—is inevitable because of the existing levels of investment, the disruption that extensive relocation would create, and the need for waterway locations for a number of the facilities to function • Existing institutions for land use planning do not afford full coverage of or control over land use decisions, such as right-of-way zoning in New York City • Existing institutions that support land use and transportation patterns are conceptually linked, but the decisions that are made are not well integrated because different agencies are responsible for them. • Existing mechanisms are in place to guide planning and development and the reconstruction of infrastructure to promote land uses that are consistent with the prevention of or reduction of the consequences of global warming associated with changes in sea level. These include capital programming, land use zoning and other forms of regulation and environmental impact assessment. At the present time, however, sensitivity to global warming effects is not directly built in as a criterion for decisions made within these areas.	• The extent to which existing infrastructure will be impaired by global warming effects, such as sea level rise • The existence and availability of technologies for infrastructure facilities that could withstand effects of global warming and adapt to the need for changes in energy utilization patterns • The ability and willingness of the region's population and economic activity to relocate substantially within the area to protect themselves from floods in the short term, but to conserve energy in the long term to prevent global warming conditions from developing; the public perceptions that would underlie such behavior. • The willingness and ability of the political system to invest in infrastructure changes in the short term to withstand global warming effects and in the long term to encourage intrastructure and land use patterns and activities that prevent conditions that lead to global warming.

TABLE 2. (*Continued*)

Categories	(a) Desired Conditions in 2020, "Goals"
Water Resources and Wastewater Management	• Delivered water supply reliable in quantity and quality for all reasonable needs, potable and nonpotable • Systems for collection and disposal of sanitary and storm sewage that meet local needs for public health and quality of life • River, harbor, sound and ocean waters that support diverse, healthy and abundant coastal and aquatic plant and animal life and a wide array of recreational and commercial activities • Water and sewer rates that are affordable and stable • Watersheds whose patterns of settlement and development support safe drinking water for all consumers, economic and social well-being for watershed residents, and recreational and scenic opportunity for residents of the Metropolitan Region

(b) Predetermined Elements, "Givens"	(c) Critical Uncertainties

• Demand management for water supply is driven not only by the economic and political costs of new supplies but also by problems relating to the expansion of the city's sewage treatment facilities
• The sources of New York City water supply—the upstate reservoirs, the Brooklyn-Queens aquifer (for Jamaica water supply) and the Hudson River (for emergency supply)—are threatened by pollution from current activities and future development. In addition, zebra mussels already infest the Hudson and threaten the reservoirs.
• Regionally, surface and ground water resources—including New York City's—will be managed "confunctively" to maximize yield and reliability
• New York City's water supply and wastewater are managed by a City agency (the NYC Department of Environmental Protection), which has powers to regulate activities and acquire land in the watersheds; a Water Board appointed by the Mayor which has power to set water and sewer rates, and a Municipal Finance Authority, appointed by the Mayor and the Governor, which issues bonds to support the systems' capital plans
• Federal and state environmental policy increasingly favors a "place-based" ecosystem protection approach: policy that
   —is tailored to the realities of specific geographic areas
   —integrates long-term environmental management with human needs—public health, economic renewal, and environmental justice
   —strives for fairness and consensus among the "stakeholders" affected by and responsible for the area
   —takes into account the relations among all environmental media—water, land and air
   —is based on sound science: priorities that are based on risk of harm, adaptive management that adjusts plans as information improves
	• The extent to which demand for New York City water supply will increase as a result of global warming effects—increased consumption (because of heat), reduced reliability of Long Island supplies (because of saltwater intrusion), increased need for Delaware Reservoir releases (because of movement of the salt front in the Delaware Estuary).
	• The extent to which the reliability of New York City water supply will be reduced as a result of global warming effects—decreased precipitation and seasonal shifts in precipitation patterns, increased evaporation, limits on use of the Hudson River (because of the salt front) and of the Brooklyn-Queens aquifer (because of salt water intrusion).
	• The extent to which global warming effects will affect the waters around the City—the Hudson River, New York/New Jersey Harbor, Long Island Sound and the New York Bight—the geometry of the waterbodies and their shorelines, the frequency and intensity of storms and the direction, steadiness and intensity of winds, the temperature and chemistry of the water, and the pattern of the currents and the combined impacts of those factors on biota.
	• Public acceptance of the water and sewer rate hikes that may be necessary to meet federal and state requirements and real regional needs.
	• Public acceptance of further demand management measures, either on a sustained or an emergency basis.
	• Public acceptance of processed water (including desalinated water) for potable and nonpotable use.

TABLE 2. (*Continued*)

Categories	(a) Desired Conditions in 2020, "Goals"
Air Quality and Health Effects	• Air quality in compliance with Federal regulations • Emission reductions accomplished without undue economic burden or inconvenience • Air conditioning universal • Reliable forecasting and timely warnings of air pollution events
Transportation Planning	• Transportation facilities that are maintained in a status of "good repair" • Lower peak-period travel congestion • Efficient, multimodal transportation system that supports economic growth and satisfies personal mobility needs • Low carbon dioxide and ozone precursors due to transportation • Restore the rail alternative for moving freight, especially across the Hudson River

(b) Predetermined Elements, "Givens"	(c) Critical Uncertainties
• The Clean Air Act and its amendments specify air quality standards for ozone, sulfur dioxide, nitrogen oxide, carbon monoxide, particulate matter and lead • Large emission reductions in nitrogen oxides and volatile organic compounds (VOC) are required to bring the New York metropolitan area into compliance with the ozone standard • The technologically easy solutions to controlling ozone have already been implemented. Ozone levels have not decreased substantially over the last two decades because of increased population and energy use. • Continued population growth and increases in energy use in areas upwind of New York City, in the absence of further controls, will cause higher emissions and worse air quality • The management of air quality in New York City will be viewed increasingly as part of a regional-scale problem • Health risks will still exist at air quality levels that meet current standards	• Political will to require expensive emission control measures and to change transportation patterns • Political infrastructure to implement multi-state solutions to regional scale pollution problems • Reduction in emissions of NO_x and VOCs from motor vehicles due to increased car pooling, use of mass transit, and electric vehicles • New technology to reduce NO_x emissions from electric generation • Continued improvements in automobile fuel efficiency and emissions • Effective monitoring program to insure compliance for motor vehicle emissions • Continued availability of low sulfur fuels for electric generation • Increased emissions of NO_x due to (i) reliance on natural-gas-fired peaking plants for electric generation and (ii) phase-out of nuclear generating plants • Changes in air quality standards • Mitigating effects of air conditioning on health effects uncertain • Uncertainty in dose-response functions for air pollutants • Number of susceptible individuals that may increase due to age or other factors
• No major, region-shaping facilities will be built. • Existing highway network • Existing network of subways and commuter rail systems • Existing transportation systems will become "intelligent" through the implementation of Intelligent Vehicle Highway Systems (IVHS) technology • Dispersed travel demand in suburban areas not easily served by public transportation • Continued use of undeveloped land to allocate population and job growth	• Growth in population and jobs • Extent to which Manhattan remains a job center • Availability of adequate financial resources dedicated to transportation • Acceptance of congestion pricing • Development of low-pollution automobiles • Extent to which telecommuting replaces commuting

TABLE 2. (*Continued*)

Categories	(a) Desired Conditions in 2020, "Goals"
Building Design and Maintenance	• Building conditions are maintained at presently accepted temperature and ventilation standards • Energy codes apply to renovations as well as new construction • Energy tariffs written to encourage energy efficiency, not just cost of service • Energy policies (such as economic development rates) that work toward higher energy efficiency, not lower energy rates • A fair value for maintaining clean air is applied to the cost of energy-consuming equipment and the more polluting fossil fuels • Exemptions for City-owned coal-burning facilities are eliminated • Qualified technical, management, and inspection personnel become available to help/push building owners/operators toward energy efficiency • Markets in pollution abatement credits are encouraged
Energy Demand and Supply	• Adequate and reliable supply of low-CO_2 energy • Effective measures to reduce peak summer electricity loads • Electric rates less disadvantageous relative to other areas • Reduction of regional contribution to global warming

(b) Predetermined Elements, "Givens"	(c) Critical Uncertainties
• Control of real estate remains in the same hands • More of the population will desire air conditioning • New energy efficiency technologies continue to be developed. • New uses for electricity and energy continue to be developed	• Extent to which new technologies (e.g., computers, industrial processes) encourage greater use of electricity • Changing government regulations and incentives for pollution control • Changes in structural technology affecting building costs • Continued competitiveness of urban congeneration due to installation costs, local regulations, and protective utility tariffs • Integration of facility "balance sheets" for energy services to minimize waste • Development of markets for trading pollution abatement credits
• Downstate electric rates are among the highest in the nation • Downstate electric utilities have peak loads in summer • The structure of the electric power industry is in flux • Short-term electricity "growth" will be satisfied by energy conservation (through demand-side management) and cogeneration (from independent power producers) • Most energy will continue to come from fossil fuels • A significant fraction of the population has no air conditioning • Intense storms will cut off electric power where power lines are above ground • New York State's potential hydropower resources are exhausted • New York State has very limited potential for renewable energy	• Continued growth in energy demands • Future energy prices • Extent to which energy conservation and efficiency improvements can continue to meet electricity "growth" • Introduction of substitutes for electricity (natural gas, "district cooling") for air conditioning • Continuation of forces producing high local electric rates • Federal requirements for reducing carbon dioxide and other greenhouse gas emissions • Extent to which "no-CO_2" electricity supply—nuclear and Quebec hydropower—becomes available and acceptable • Further structural changes in the electricity supply industry • Whether New York Power Authority hydropower will continue to be essentially restricted to upstate users • Continued growth in personal transportation • Whether an economic low-CO_2 automobile can be developed

Report of the Scenario Planning Group for No Climate Change
"Little Green Apples"

Facilitators

R. LAWRENCE SWANSON

Waste Management Institute
Marine Sciences Research Center
State University of New York at Stony Brook
Stony Brook, New York 11794-5000

SHELDON J. REAVEN

Department of Technology and Society
State University of New York at Stony Brook
Stony Brook, New York 11794-2250

Rapporteurs

ANNE MOONEY

Waste Management Institute
Marine Sciences Research Center
State University of New York at Stony Brook
Stony Brook, New York 11794-5000

SALLY BOWMAN

Stony Brook, New York

R. LAWRENCE SWANSON: Our group was assigned the task of planning the scenario for "no change due to global warming." The only really noteworthy change is that there will be an increase in sea level of about 4 inches. That's the historic rate over the last century in the New York metropolitan area. It is based on such things as isostatic readjustments, warming of the oceans, melting of the glaciers, and so forth. This is in fact the reality that we have been living with over the past century.

The group really looked at that four-inch rise and said that they did not recognize many significant effects as a consequence over the next twenty to thirty years. Possible exceptions are that there will be a continued loss of wetlands, particularly wetlands that are in front of areas that have been bulkheaded. So that the wetlands will not be able to retreat or move shoreward with the progression of the rise in sea level which they would tend to do naturally if there were not hardened surfaces behind them. The consequences of that, perhaps, is that we will continue to lose fishery resources, particularly those resources that need to have a wetland environment for nurturing, growth, maturity, *etc.*

After six or seven hours of debate, the group was a little bit concerned that the amount of rise in sea level that we were talking about was so difficult to visualize that, in fact, they really did not appreciate what four inches might be. And so, one of the recommendations is that we re-look at just what a four-inch rise in sea level might mean in terms of infrastructure, *etc.* We suggest limitations that may be placed on the problem due to climate changes and sea level changes that might occur *not* due to global warming. I think there is an overall desire to look more in depth at what the consequences might be that we perhaps are giving short shrift to. For example, there is the problem of more basements in the New York City area that might become wetter as a consequence of the change in sea level and saltwater intrusion. Areas in Brooklyn might see consequences that we have not given enough credit to.

We realized that looking at the change in sea level would provide an opportunity to consider a number of social, political, and environmental problems that are confronting us today and will continue to confront us and perhaps will be exacerbated to some degree by the very small rise in sea level over the next 30 to 40 years. I'd like to go over some of those with you.

We thought there needed to be a very broad design vision of the waterfront. I think we heard at lunchtime that there are such plans. But in our vision, we thought that there ought to be mixed modes. We just can't have the waterfront be a greenway for bicycle paths. There are still very important functions of government and society that need to take place at the waterfront. As a consequence, there will have to be mixed use. For example, we know that sewage treatment plants in twenty or thirty years are still going to be on the waterways. That was a given in our particular scenario. To think that we are going to have greenways going through the sewage treatment plants may be somewhat unrealistic. We also know that there are areas today that are perhaps swimmable in New York City, and we would want to try to maintain that water quality for that type of use in certain sections of the city. So this whole notion of mixed use we thought to be very important.

We wanted to review the role of offshore islands in the context of waterfront development. There is perhaps a great deal that can be gained by using offshore islands for a number of purposes, purposes that are perhaps not going to be as tasteful to us because of the commercial or waste disposal problems that the city is confronting. Perhaps offshore islands can serve a function in that capacity. There has been talk over the years, for example, of using the Hoffman-Swinburne Island system in Lower New York Bay as a mechanism to dispose of contaminated dredge materials. Or perhaps just build those islands up to use for the purposes of some other activity, perhaps using them for garbage and trash that the city can't otherwise handle with existing facilities. So doing things that are innovative with offshore structures may be a very important consideration.

We also think it is important that the harbor be maintained as a working harbor. We want to have a clean harbor, but we also think that the economic vitality of the port is very much dependent upon the fact that the port is still a port. It may be somewhat unrealistic to think of the port as a place that is going to be the most desirable place to swim and fish in years to come. Even though those are goals—to have fishable, swimmable water—I think it is very important to have a working port for the economic vitality of the area.

We thought that we should take advantage of the existing port facilities to encourage new high-technical growth. For example, the port might serve as an area where people would come if perhaps an offshore airport were constructed. That has been discussed in the past: move Kennedy Airport out to sea, increase the capacity of that particular airport, and have a very rapid system of people movement directly to downtown Manhattan. That kind of innovation may be important.

Deep-water ports are still extremely important. One of the problems that confronts New York City in particular is that they have to dredge on a continuing basis to keep the port and harbor open. In fact, on numerous occasions, the very nature of the port as a port facility has been threatened because we have not been able to dredge owing to the contaminated sediments that exist in the harbor. So we might want to look at how to accommodate better deep-water facilities in the outlying areas as opposed to in the harbor itself.

We may also want to look at high-tech recycling ventures in the New York metropolitan area. Recycling was mentioned as an issue at lunch. Let me point out that from an economics point of view that waste materials right now are one of the most important economic commodities that are going in and out of the port of New York. Nowhere in the country is there more of a waste stream that can be made into a legitimate commercial venture than right here in the New York metropolitan area. Not only do you have a very captured market in New York City, but you also have the markets that come from Long Island and New Jersey. There is a lot of money in doing something innovative in the recycling arena.

There was also the suggestion that all new economic vitality in the area is not going to be water-related, that inland activities are extremely important. However, the waterfront offers an opportunity to make those inland commercial ventures more attractive. The reasons why you might want to come to an area is because it is on the coast. It has beaches within a reasonable drive. It has the opportunity to go boating, and so forth. The excitement of a city on the water may offer opportunities to ventures that are not necessarily directly tied to the water.

We thought that there ought to be environmental accounting. Perhaps some credit, tax credits or other kinds of credit associated with the encouragement of "green" building. Battery Park is a good example of that kind of thing, where there has been some thought of how parks and so forth associated with that venture could make the area more attractive and more usable to the population of the city.

Time and time again in our discussions, it was pointed out the clean, drinkable water was of the highest priority. We are talking about the drinking water supply. As a goal, as an objective, we think that this is about first on our list. That involves a lot of things as to how we manage the water resource upstate. It also makes it imperative, perhaps, to look at the notion of water conservation. As we discussed a little bit yesterday, water conservation is important not only with the idea of making the water supply last longer and keeping expenses down. But once you get that water to the home, it then has to be redistributed and at a cost that can be avoided if you conserve water in the first place.

A plan to design and implement a program of routing infrastructure in the public right-of-way is another idea that I think is particularly important. Almost everyday you pick up *The New York Times* or other local newspapers and read that a roadway is being dug up in order to get to a water main that's a hundred years old, or there's a sewage pipe failure, there's a storm water this or a storm water that. This is a very expensive undertaking, but some consideration should be given to rerouting electrical lines, water delivery systems, sewer systems, storm water systems, in order to make the overall process a lot more effective, a lot more clean in many respects, and perhaps a lot less expensive to deal with than current practices.

We thought that there ought to be a long-term, broad approach to energy, air quality, and transportation. It seems at least to some of us that these activities very often are studied in a vacuum. Transportation is considered separately from energy, energy is considered separately from air quality, and really they need to be linked. The debate comes up, for example, in the transportation issue of congestion. Congestion is a cause of degraded air quality. Some people would suggest doing away with automobile emissions by going to an electric car. But the repercussions of that are that power has to be generated somewhere else to supply energy to an electric car, and so we think we ought to look at these air quality, energy, and transportation systems more holistically than perhaps has been done in the past.

There is a considerable investment already in the New York metropolitan area in infrastructure: sewage treatment plants, bridges, *etc.* The first things that are typically cut in budgets of government are operations and maintenance funds. There should be a concerted effort to assure that operations and maintenance funds continue to be made available in order that capital investments that have already been made are kept up to snuff. A great deal has been made in recent months about the poor quality of the bridges in New York City. Whether they are or not, I don't know. But nevertheless if money had been pumped into the system over the last fifty or sixty years to keep those bridges up to snuff, we might not be faced with some of the problems of coming up with more capital money now to replace some of the facilities.

Better education in long-term planning is also probably very important. You have to remember that the political process usually is very short-sighted. It's as long as the term of office of the person who is sitting in the office. That's just the facts of the way our government happens to run. We believe that we have got to stress more with the general public the necessity to look at these problems over the long term, and to get the public to begin to think in terms of twenty, thirty, or forty years. In fact, if the public begins to think in those kinds of terms, then perhaps our politicians will be able to respond.

In the air quality area, there was a belief that we need to use greater public transportation. We need to encourage the concept of work-where-you-live, or live-where-you-work, and stop the commuting. We must examine other things that are contributing to air quality problems besides transportation. It was mentioned that there is the transport or advection of contaminated air into the area from other areas; what is the cause of that? What can we do about that? How can we work with other regions to help solve a regional problem as opposed to just looking at the automobile locally?

Finally, we thought that there was a particular problem in New York in looking at contaminated waterways. We have contaminated waterways that literally we can't use anymore because we can't dredge them in order to maintain them for any useful purposes. These are indeed, for all practical purposes, hazardous waste sites. The city must begin to look at these contaminated waterways and bring them back into use because the shoreline is in fact a valuable resource. We would like to see more and better utilization of that resource in years to come.

I think that fairly well summarizes what we had discussed. Sheldon, I'll ask you first if I have left things out, and you can make corrections.

SHELDON REAVEN: There are one or two areas where you might want to add a point or two. One overriding theme of our scenario—no change—is that it opens the coastline and the waterfront to use. We presume that in the other scenarios where you have a rising sea level that you have to barricade or create barriers to use and move current activities back from the beaches. So whether we are talking developed waterfront, or beaches, or other kinds of shoreline, if we are not going to talk about rising sea level, our overall theme was we are going to open things to use.

One of the major areas that we think is quite promising might be grouped under the rubric of integrated aquaculture or fishery-type systems. This would involve restoring and sustaining the fisheries in various ways, for example, using waste heat from power plants to warm kelp beds, or in connection with shellfish farms of various sorts. There are a number of very interesting schemes. This would provide, by the way, relatively low-skilled jobs that will be in demand in the area. It will create a sense of bustle and activity on the waterfront. It will be environmentally benign. It will also confer, for example, Clean Air Act benefits on the utility through the use of waste heat, also through possible credits through recycling of the shellfish shells that are processed into various other products. That's just one example.

Another major fish-related theme was a "fishport," a proposal that had been bruited about in New York City but for some reason or other was not implemented. I don't know the history of it myself, but we all found it quite interesting, and we thought that the fishport proposal deserved to be resuscitated.

The only other point I would add is that we did discuss the necessity of improving beach access, and waterfront and coastline access generally, and we thought that this required in some cases a radial approach, that is, the access would come from inland to the individual points or nodes of recreational activities in our mixed-use plan, some commercial, some recreational, and so on. In other areas, the preferred form of public access would be the linear, belt bikeway or greenway or whatever. In some cases, that might be appropriate, but one could not generalize overall, and there are lots of areas where the radial approach would be superior.

DISCUSSION OF THE REPORT

ROBERT ALPERN: Just one overall point. Our scenario was: no impact from global warming. But that didn't mean that in fact current climatic conditions would prevail in the year 2030 or 2070, whatever the benchmark year would be. The fact is that the historic record shows much more extremes of climate, forgetting about global warming. There is a tendency in all of these discussions to be complacent about where we have been for the last fifty years. Forgetting about global warming, we are going to see real possibilities for extremes that we haven't seen in our lifetime.

REAVEN: I'm glad Bob brought that up, because that was a strongly emphasized point in our considerations. The scenario assumption of no global change does not imply that there will be no effects. There should be increased variability in the frequency and length of storms and other weather conditions. We have essentially been through a lucky period of relative stability during the last half century or so. For reasons independent of global change, things won't be that stable. So we should be prepared for that in any event.

ANTONIA BRYSON: You implied, but I'd like you to confirm, that it was your conclusion that even though there is going to be an albeit small rise in sea level, there will be no impact on existing either industrial or recreational waterfront usage.

SWANSON: The feeling was that over twenty or thirty years that it would hardly be noticeable, that if you raised the water level four inches and the storm surge was four inches more than it was, say, in the December storm of 1992 that there would still not be much more than was noticeable at that particular time. I am not saying that all of us agree with that; that's just a consensus of the group.

JEANNE FOX: On the water supply front, did you get into a discussion at all? I was involved with the Delaware River and the Delaware River Basin Commission for three years. All the reservoirs are interconnected. New York reservoirs are needed sometimes at times of potential drought to increase water flow in the river so that southern communities—Philadelphia and south Jersey—have water supply. There is already intrusion going on. Did you have any discussion of that? Because clearly there is something now where four inches would make a significant difference.

REAVEN: Yes, we did discuss that, guided by the valuable insights from Bob Alpern. We discussed the issue of possible interwatershed exchanges between the Delaware, the Hudson, and the Long Island aquifer due to differential water supply and water demand, for example, as you mentioned. The consensus of the group was, for our scenario, there would not be any significant change in that regard.

FOX: I'll have a lot to say about that later then.

PHILIP JESSUP: Did you have any discussions of how energy considerations and transportation problems and issues could be better integrated? Traffic engineers these days are preoccupied primarily with congestion, and there are probably very few who could tell you anything about energy. How could we get them together?

REAVEN: There are a number of people in the room who do have extensive experience in this area. Let me just think about it, and maybe later I'll give you a better answer. We did discuss those connections, and we did say very clearly that transportation planning had to take into consideration changes in energy uses, liquid fuel sources, electric vehicles, and the location of transportation facilities.

JESSUP: If you find any solutions, I think it is worth a million bucks.

REAVEN: Well, maybe I'll think a little harder.

Report of the Scenario Planning Group for Medium Climate Change
"Apple Fritters"

Facilitator

WAYNE TUSA

Environmental Risk & Loss Control, Inc.
309 East 90th Street, Suite 4
New York, New York 10128

Rapporteurs

PHILIP A. CHIN AND SHINO TANIKAWA-OGLESBY

Marine Sciences Research Center
State University of New York at Stony Brook
Stony Brook, New York 11794-5000

The "Apple Fritters" scenario planning group was tasked with developing recommendations that would increase the likelihood that New York's infrastructure would continue to function given the effects of climate change resulting from a medium greenhouse effect. The prevailing conditions for a "medium" climate change are given in TABLE 1.

The challenge to Apple Fritters was to identify positive and negative impacts on the existing infrastructure due to medium climate change and identify recommendations that would alleviate, ameliorate, mitigate or eliminate the predicted impacts while meeting the 'Desired Conditions' identified as part of the scenario-planning process. The Apple Fritters also provided a number of suggestions which slightly modified the set of Desired Conditions (TABLE 2). A summary of the findings of the workshop is as follows.

DEMOGRAPHIC AND ECONOMIC FORECASTS

As sea level rises the populace currently living and working near the shoreline may have their activities constrained. We don't know what percentage of the populace may be affected, but the group did note that some low-lying areas will be more markedly affected than others, especially during storms. We expect the populace to shift out of immediate harm's way. However, to where the population (transients and immigrants) will shift is another unknown. Discussion centered around potential additional suburban sprawl or whether New York City would become even more vertically developed.

The group understood the importance and dependency of the economy on infrastructure and that a deterioration of infrastructure would lead to worsened economic conditions. We also understood that the economic burdens resulting from global warming, such as flooding, damaged transportation systems, and

TABLE 1. Assumptions for Medium Climate Change Scenario

Global Average Temperature	
2030	+2.0°F
2070	+4.3°F
Local Temperature	
2030 Winter	+4.0°F
Summer	+2.7°F
2070 Winter	+8.1°F
Summer	+5.4°F
Sea Level Rise	
2030	+11 in
2070	+28 in
Precipitation	
2030 Winter	+8%
Summer	−8%
Summer Soil Moisture	
2030	−18%
Frequency of showers and thunderstorms	—more—
Day-to-day and interannual variability of midlatitude storm tracks	—less—
Tropical storm frequency and intensity	—more—

reduced water supplies would be locally specific, *i.e.*, impacts might be more costly in urban Manhattan than rural Fire Island. The cost of living and of doing business would likely rise for affected communities. The group did not have the pertinent information and was unable to quantify the incremental negative economic impacts. Cost recovery and direct and indirect costs were mentioned as issues of importance. Mitigation measures suggested by the group were:

- Tax and financial incentives
- Increased energy efficiency

TABLE 2. Changes to the Desired Conditions Suggested by the Apple Fritters

Demographic and Economic Forecast
- Improved quality of life
- Increased economic growth without significant increase in resource consumption

Water Resources and Wastewater Management
- Federal and state environmental policy increasingly favors a 'place-based' ecosystem protection approach (moved from *Predetermined Elements* category to the *Desired Conditions* category)

Air Quality and Health Effects
- Adequate protection of public health
- Universal air conditioning (deleted)

Transportation Planning
- Reduced travel demands per capita

Energy Demand and Supply
- Reduced energy use per capita
- Increased use of "green" power

Other Recommendations
- Improved environmental awareness
- Implementation of community based emergency response mechanisms

REPORT OF THE "MEDIUM CLIMATE CHANGE" GROUP

- Construction of protective structures to prevent flooding, *e.g.*, bulkheads, dikes
- Development of regulatory mechanisms to prepare for these impacts. One potential approach would be to discourage inappropriate development along the shoreline

WATERFRONT PLANNING

There were expected to be both positive and negative changes in land use patterns near coastal areas. Especially hit hard may be the beaches of Long Island, New Jersey and other low-lying areas. The most significant impacts would be within these areas, particularly on the tourism and real estate sectors. Property values are likely to decrease, and businesses may relocate, possibly out of the region. The pace of new development and redevelopment may slow. However, development projects that do take place will be those that take advantage of the new character of the area. The group recommended that:

- Global warming issues be incorporated into land use planning activities
- Building codes be modified for both new and existing buildings
- Regional waterfront land use planning incorporate waterfront protection as appropriate
- Development projects incorporate impact assessment evaluations

LAND USE AND INFRASTRUCTURE PLANNING

The group expects limited impacts on land use patterns in noncoastal areas, but the intensity of land use throughout the area may shift. For example, the population demands on existing infrastructure within the area will change as people and businesses move to higher ground. The cumulative effect may be a shift in real estate values. Discussions also took place on the topics of land use zoning, the efficiency of cities versus suburbs, and discouragement of low density sprawl. The extent to which climate change might impact agricultural production was also discussed. Recommendations for mitigation included:

- Incorporation of global warming issues into land use planning activities
- Modification of building codes for new and existing buildings
- Revitalized regional planning efforts

WATER RESOURCES AND WASTEWATER MANAGEMENT

Increased saltwater intrusion was predicted to lead to reduced local groundwater supplies. In the catchment basin, drier summers will also result in reduced water availability. Winter flooding may be more frequent owing to the increased frequency of precipitation events. Associated with this will be increased costs for pumping, storm water and sanitary wastewater management. The expected rise in sea level may also have physical impacts on the performance of wastewater treatment plant outfalls. Recommendations for mitigation included:

- Increased water conservation and reuse
- Increase pumping capacity
- Increased efficiency of the water transmission system

AIR QUALITY AND HEALTH EFFECTS

The group predicted that there would be limited impacts on air quality which are largely secondary impacts associated with transportation and pollution control activities. For example, increased acidity of atmospheric deposition could result in a limited increase in the corrosion of buildings and other structures. In the health area, higher temperatures and reduced air quality in the summer may lead to a predicted additional five hundred deaths per year (Kleinman and Lipfert, this volume). Hardest hit will be the young and elderly in lower socioeconomic classes. No mitigation measures were suggested for this category.

TRANSPORTATION PLANNING

The increased frequency of precipitation and storm events would likely give rise to more frequent flooding of surface routes and subsurface transportation systems. The vulnerability of the local airports, particularly on Long Island, was discussed. The group noted that corrosion damage resulting from seawater is typically more extensive and severe than from freshwater. Overall, there was the potential for substantial deterioration of the transportation infrastructure. Recommendations for mitigation included:

- Selected physical protection of surface transportation systems
- Improved flood protection at vulnerable points of the subsurface transportation systems
- Improved pumping capabilities
- Improved storm water management
- Improved flood protection at local airports

BUILDING DESIGN AND MAINTENANCE

The group agreed that there will be an increased demand for air conditioning. Recommendations for mitigation included:

- Improved energy management standards in building codes for new and old buildings

ENERGY DEMAND AND SUPPLY

The group agreed that there would be an overall increase in energy demand. Recommendations for mitigation included:

- Increased energy conservation efforts

- Increased use of energy sources that do not release carbon dioxide
- Evaluation of the supply and transmission of energy on Long Island

OTHER RECOMMENDATIONS

In this catch-all category the group came up with a number of additional recommendations:

- Increase education and awareness for consumers
- Increase education and awareness for decision-makers
- Development of incentive based systems
- Completion of energy and emissions inventories
- Implementation of a research program on potential local global warming impacts
- Development of quantitative data on mitigation alternatives and costs
- Integration of global warming issues into all infrastructure planning efforts

CONCLUDING REMARKS

The group felt that they did not have enough information on the topics discussed and that as a consequence the group may have underestimated the potential impacts on the region's infrastructure, particularly secondary and ripple effects. The group also indicated that much more assessment and analysis of the potential impacts of global warming on the region's infrastructure was needed and that implementation of an appropriate set of mitigation measures would likely be particularly challenging owing to political and economic constraints.

DISCUSSION OF THE REPORT

RITA MEYNINGER: I have some comments. The agency I am with, the Federal Emergency Management Agency (FEMA), does a lot of predictive work based on the need to prepare for disasters, exercise the plans, and then respond to them when they happen. So we have to conceptualize a lot of the things you are talking about. In Region 2—I understand Rae Zimmerman knows about and alluded to our hurricane model—the hurricane effects model for downtown Manhattan would have this building with 30 feet of water in front of it in a Category 3 hurricane, which is not extraordinary. I think we used existing water levels in our model and not the additional rise in water level that your scenario encounters. But we used a combination of adverse things like a storm surge. Most of the streets of downtown Manhattan would be under water. It is hard to think about, except that you have all seen on national television roads in France and in Holland with the cars coming down the road in flushes of storm water. That could exist here.

The next impact of such a thing is that the subways are closed; they're flooded. The Lincoln Tunnel, the Holland Tunnel are closed. The access to all the bridges are under water. And people, in order to evacuate, move up. They don't go anyplace else; they move up the building. So these are not unusual effects. And, of course, we may not have electricity. We may not have communications equipment. This is not an unusual aspect.

The Northridge earthquake gave FEMA a very clear picture of the kind of unexpected event we know is going to occur. I mean, unexpected as to where it is going to occur and what the severity will be. One of our planning programs in FEMA is earthquake hazard planning to anticipate the next earthquake, which we assume will be reasonably soon, and its severity. We plan for the most severe catastrophic events, and we have the plans in place that are exercised constantly to respond to it. A lot of the things you have here are things that are incorporated in various elements of our planning. Rae Zimmerman said she used some of that in her talk yesterday because she is familiar with it.

We are now working on a new element of FEMA. I called the president of the American Society of Civil Engineers this week to inform him and other engineering organizations about it. Under one of the new aspects of the legislation that funds FEMA, we plan to spend a lot of money to do mitigation. The dollars for mitigation are enormous. There are going to be some positive effects from this. For example, FEMA estimates the cost of the earthquake in Northridge, one county in California, to be about $30 billion. The new legislation requires us to spend, in addition, a specific percent of a certain part of the cost of that $30 billion. The total money that we will spend on mitigation is almost a billion dollars in that one county. That's going to exist all over the country now under this new legislation. We are currently augmenting our ability to provide mitigation plans to have them on the shelf in the state emergency management organizations that we, FEMA, fund in all fifty states, Puerto Rico, and the Virgin Islands.

Many things are part of the mitigation efforts, like property acquisition where you move people out of flood plain areas. Some of the real effects that can be anticipated over the next years may be mitigated to a certain extent. It's an enormous program, and we don't know what the dimensions of it will be. But I think it is going to help if we do have a significant rise in the effects of global warming. We are going to have some ability to mitigate it.

RAY RUGGIERI: Jeanne was saying that she is glad she is moving to the 20-something floor, and I am just sitting here trying to figure out whether I should be happy or worried that I am on the 82nd floor.

JEANNE FOX: You'll have a lot of company.

RUGGIERI: Yes, but they'll never leave. You mentioned about population moving away from the waterfront. Did you think at all about where they are going to go? South Bronx, South Jersey, South Carolina, Idaho—I'm serious. What I am getting at is what does this portend for how the whole country looks, and what does it mean for New York?

WAYNE TUSA: We struggled with that. We had two responses. There might be some prudent residential owners and business owners who might choose to move in advance. We thought more likely what would happen is that most would just stay and weather it. We didn't think the impact numbers would be large. When we asked ourselves whether they would move locally—three blocks inland—or move to Ohio, we thought they would stay locally.

DOUGLAS HILL: Moving upstairs, you never considered.

TUSA: No, we hadn't thought about that one.

PHILIP JESSUP: Having been a member of the second panel, I just want to ask Rita a question. Were you saying that we may have underestimated the potential impact of storm surges and flooding on New York City streets and subways and so forth?

REPORT OF THE "MEDIUM CLIMATE CHANGE" GROUP

MEYNINGER: You yourself said that you don't have any numbers. FEMA needs to know how many people we are going to put in the field, and what kind of talents they should have, and where they are going to be. We send teams out two days before a hurricane lands, and we have to know where it is going to land. So we are into predictive modeling. I think you have underestimated some things because you simply don't know, and we don't have all that kind of information either. But some things that are going to start coming in line in the country under the guise of mitigation are going to reduce the adverse effect.

And where would people move? We don't know where they'll move. I would hope that there is a sociologic study done of the Northridge population. Do people really leave? Because we can never get them to move away from the shoreline.

TUSA: I just might add that Rita and I have talked about this before. If I have understood you right, Rita, mitigation includes things you do before the event occurs.

MEYNINGER: That is it, exactly. But the money doesn't come until the day of the disaster.

FOX: Maybe that's one of the things that we need to change. Two questions: the question about the air conditioning, in all seriousness. Are they all going to give it up for the whole year, or just for the wintertime? Secondly, the question about waterfront land use and protection planning. Can you go more into what you meant by that? In New Jersey they have a coastal area facility review act where there is some planning and special protection and requirements for the coastal area. Is that the type of thing you mean?

TUSA: Yes, in response to the second question, that is exactly right. The kind of planning that is going on at the local level, our suggestions would look at that regionally. Include impacts from global warming. On the first question . . .

JESSUP: I should probably answer that since I am the one who suggested it. I suggested that universal air conditioning would significantly increase energy demand in this region. It would also contribute significantly to the heat island effect which will cascade. And having lived in New York for ten years in Brooklyn and never used an air conditioner, even on the few hot summer days that we had out there, I just didn't think that air conditioning was something that everybody needed. In some places in New York, yes, but I think that a lot of the houses in the boroughs and so forth, where there is more breeze . . .

FOX: Well, if you are near the water and can get a breeze. I don't use the air conditioner very often myself, but I see it primarily as an environmental justice type of issue. We all know people who have air conditioning on all year. Everybody knows people who have air conditioning on from when it is seventy-two degrees out. We need to get to those type of people. But we also know a lot of people can't afford air conditioning, maybe based on the building construction or the electrical situation, because of money, and they don't have air conditioning when it is ninety-five degrees. That can be a health problem for large classes of people. More is needed than just making a general statement, because what it is saying is: we who have it can keep it, and we can use it as much as we want.

There is an education element, obviously, but a lot more thought has to go into it than just eliminating that as a goal. It is easy for me to say because I can turn it on when I need to, and ten times a year I do it. The house I have down on the shore, although it is twelve feet above sea level, will probably be gone with eleven inches more water under the Apple Fritters scenario. It's something that we really have to think more about.

JESSUP: We did say in our group that the market will determine who buys air conditioners or not. What we were saying was, we didn't want Con Edison going out and buying an air conditioner for everybody in the city. Those populations in the city that are vulnerable, the elderly and so forth, who are extremely vulnerable to heat, should have air conditioning. We just didn't want everybody to have it.

ANTONIA BRYSON: Following up on that, I was wondering if anyone did consider in that connection any kind of natural cooling, either in land use or building design principles, whether one of the impacts would be to shift towards more natural cooling principles. And in that connection, also to make the comment broader, it does strike me as odd that you considered that there would be no impact, other than in the coastal areas, on the way people lived. And I was wondering if there was any discussion about inland areas. The increased heat—maybe it wasn't significant enough—affecting the type of environment that people wanted to live in, again either the greater environment in terms of the natural landscape, or the built environment in terms of whether they live in high-rises or houses or whatever?

TUSA: I think that the analysis that we went through where we restricted the impact issue to near-coastal had to do predominantly with waterfront development. In all the other arenas, we were thinking about the region in its entirety; I guess that wasn't that clear. On the air conditioning issue, what we really did is take it out of the category as a goal—everyone has an air conditioner—and put it back into the marketplace and let individuals choose. That's really the only thing that we did. We didn't say air conditioning was not a good idea. We said, instead of it being universal air conditioning as a social goal, let's take it out of the desired conditions goal category and put it back in the marketplace and let consumers choose.

BRYSON: I would say that is even looking at it from the wrong side. The goal has to be a comfortable level of internal and external temperature for the population. Whether you achieve that through air conditioning or through some other mechanism is irrelevant.

RUGGIERI: I've got two questions on regional planning. One is, could you define the region? And the other one is—and I am not being facetious—could you define what "reinventing and doing regional planning" means? I say that because I submit that we do regional planning. We just don't make regional decisions.

FOX: He should know.

TUSA: I think that we are in complete agreement.

RUGGIERI: OK, but what is the region?

TUSA: The region was the 31 counties, all the coastal counties, New York, Connecticut, Nassau, Suffolk, and all of New York City.

RUGGIERI: OK.

PHYLLIS Y. ATWATER (*New York City Regional Director, New York State Department of Environmental Conservation*): I was a member of the group, and I guess the one thing we came to overall was that the assumptions and the analytic framework that we were engaging here could be used more broadly than it currently is, whether or not one assumed that there will be global warming. The whole exercise, if you will, illustrated the extent to which planning is not effective or the extent to which plans are not joined currently. Or the lack of accumulation of our understanding of the impacts of things, whether it is the inventory of our infrastructure across this region because we could look that way, or think that way, or make policy that way. So that regardless of the assumptions—and in response to the representative from FEMA—although we couldn't quantify anything, the analytic process had to go through all the same steps I am sure you go through. We need to do that more broadly in every kind of infrastructure planning. We need to be ready for these questions.

MEYNINGER: One thing that was on Wayne's list: I was amazed to learn of the value of the education of consumers. We found in response to both the firestorm and the earthquake in California that there were examples of people who actually bought the booklets we put out—individual homeowners—and who applied everything that was in the booklet. I don't know if either homeowner spent more than a thousand dollars. One earthquake-proofed his own home. His was the only house standing in blocks. Not a window was broken. Not a wall was cracked.

A civil engineer bought the booklet and fireproofed his home. And the same thing. So we totally underestimate the value and application of educating the consumers. We don't do it in the public schools. We have to relook at the methods we use. We tend to look at the large pieces: regional planning. And then we don't do it. Maybe we ought to start going backward and look at the small pieces and how effective they may be.

MALCOLM BOWMAN: That reminds me of the Biblical story of Noah and his family. He tried to tell everybody that a great flood was coming, and no one listened. They thought he was crazy; he started building the ark. And then it rained for forty days and forty nights, and he and his family and the animals were all that survived.

One other comment: some of you may be thinking: eleven inches, a puny eleven inches! How could eleven tiny inches make all this difference? Well, the point is that it's on top of everything else. If we have a major storm that occurs during high tide during a spring tide, the net eleven inches can make all the difference. Every ship that sails the sea has a marking on the sides called the Plimsoll line. That tells the captain that he shall not load his ship any higher than that line when the ship moves into the water. Before we had that Plimsoll line on all ships, there were all kinds of tragedies during storms at sea because the temptation was just to load up. But the Plimsoll line has saved countless lives and countless ships. And those eleven inches on any ship can make a big difference.

Report of the Scenario Planning Group for Accelerated Climate Change

"Apple Crisp"

Facilitator

SAMUEL C. MORRIS III

Department of Applied Science
Brookhaven National Laboratory
Upton, New York 11793-5000

Rapporteur

MARTIN H. GARRELL

Department of Physics
Adelphi University
Garden City, New York 11530

THE SCENARIO

This group was assigned a scenario for accelerated global change, a scenario characterized by a temperature increase of 4.5°F in summer and 7° in winter and a rise in sea level of 19.5 inches by 2030. By 2070 the temperature rise for the scenario was 9° in summer and 13° in winter, and the projected sea-level rise was 4 feet. A map provided by one member of the group suggested that the four foot increase in sea level would produce water lapping at the entrance to the World Trade Center, the site of our meeting, and would flood the first story regularly. In addition, according to the scenario, winter storms, including Class Five events, e.g., the 1938 New England Hurricane, would become more frequent while overall precipitation would be reduced; coastal destruction and regional water shortages would thus become highly probable.

We assumed that the conditions given for the year 2030 would build steadily and that impacts might be felt well before 2020. There was some discussion that Manhattan might not be tenable as the economic center of the region, and, consequently, that major commercial and financial institutions would be forced to move to higher ground. The general consensus of the group, however, was that social and commercial forces would combine to do whatever was necessary to maintain Manhattan. Changes and dislocations would occur because the entire coastline of Long Island and New Jersey could not be protected. Property would be lost, and massive investments would be required. The effects would be comparable with those from a major earthquake, perhaps worse. Moreover, New York City would not be alone; every coastal city would suffer similar impacts, and inland areas would face problems of their own.

Unlike the events following an earthquake, the resources of the entire country would not be focused on the impacted region. Because of the gradual onset of the scenario over 30 to 80 years, however, there would be time to prepare and time to act. If thousands or millions of families and businesses lose everything at once, the situation is a disaster requiring immediate and massive action. If, on the other hand, those same families and businesses understand that they must move or make investments to protect their property over the next 50 years, they should be able to cope. The overall sense of the group was that the viability of the city and the metropolitan area could be maintained.

Many of the ideas and suggestions developed by the group would be rejected out of hand in today's political climate as infeasible because of costs. Impact assessments required for tide gates, dikes, and pumping stations would also thwart planning. Given the threat of severe effects of climate change and the expectation that, under the Apple Crisp scenario, such a threat would be evident early, the group felt that the rules of the game would change. A demand would come from many quarters to take action. Indeed, one political problem might be the prevention of panic and ill-advised haste.

MITIGATION AND ADAPTATION

To *mitigate* global warming, reducing or postponing its impacts, emissions of greenhouse gases would be a first-line strategy in the metropolitan area and in the nation at large. Global efforts at mitigation would be strengthened here. Federal and State governments would mandate mitigation and legislate taxes or other measures to assure the reduction of emissions. General practices of construction and operation would change over time to reduce emissions. These measures would not be unique to New York City, nor, under the scenario, could they be expected to have any significant impact in reducing the effects of global warming on the metropolitan area. They might play a part in preventing conditions from getting too much worse after, say, 2070. Expected measures would be imposed in a similar fashion to those required regionally by the Clean Air Act.

Because of the constraints of time, we decided to focus on *adaptive* rather than *mitigative* measures, considering how the region would respond to protect its economy, transportation network and infrastructure. We therefore discuss these adaptive measures in detail in what follows.

TRANSPORTATION

The New York metropolitan area already faces a future that includes increasingly stringent constraints on its transportation network brought about by the age of the system, the limited flexibility due to limitations on available space, and the requirements to meet ozone standards. Greenhouse mitigation measures will add to such constraints but may, at the same time, help to push the system to reduce emissions of other air pollutants, including precursors of ozone.

Sea-level rise will further aggravate transportation patterns. Group members described potential impacts on coastal highways, based on experiences in Class Three storms, *e.g.*, December 1993. The combination of sea-level rise and more frequent and more energetic winter storms would require that the Belt Parkway, FDR Drive, and parts of the Garden State Parkway be protected by dikes, raised in elevation, relocated to higher ground or abandoned. Even before this can be accomplished, storm flooding will close these major highways more frequently for short periods. Increased flooding of roadways may also occur in inland counties of New York and New Jersey as a result of more severe winter storms.

Increased disruption of major highways in the city, combined with concurrent pressure to reduce ozone and carbon dioxide emissions even further than today is likely to lead to greater emphasis on mass transit for travel into Gotham. Indeed, we speculated that increasingly stringent efforts to reduce ozone may lead to a ban on all but zero emission vehicles (ZEVs) in Manhattan before 2010, even without global warming. Improvements and expansion of the rail system will be required. Even the mass transit system will not be immune to the effects of global climate change. Coastal flooding, increased winter precipitation and more severe winter storms will increase flooding in subways and rail tunnels. Flooding and mud-slides ocasionally disrupt suburban rail service even now. Investments to protect stations and ventilation systems from flooding, including better pumping systems, will therefore be required. Improvements to tracks and rights-of-way on suburban rail lines will be needed, too.

In the period of flux around the turn of the century, when there may be considerable uncertainty about the viability of the coastal location of new highways or rail lines, more emphasis on waterborne transit may be in order. Increased ferry service to Manhattan from Brooklyn, Staten Island and New Jersey may prove ever more cost-effective as the years go by.

WATERFRONT

Initially the group focused on New York Harbor. Sea-level rise would threaten valuable real estate in Manhattan, Brooklyn, Staten Island and New Jersey. Increasing risk of storm damage and other consequences of global warming would make New York City less desirable. The commercial center and much of its population might shift to a safer location across the rivers or simply be displaced by other urban centers in the northeast that were less impacted. This seemed unlikely from both commercial and political viewpoints.

Given the sea-level rise posited in the scenario, tidal gates, dams and locks could theoretically be built across the Verrazano Narrows, the Arthur Kill, and Hell Gate which would protect the entire harbor. But, unlike Amsterdam, New York has the Hudson flowing into it, so it would be necessary to pump the entire flow of the river over such dams to prevent the city from being flooded by the river itself. Alternatively, the bulk of the flow of the Hudson could be diverted far upstream, possibly down the Delaware. The first option appeared impractical and unreasonable while the second option appeared to be beyond the ability of the group to address intelligently. Consequently, the group went no further with these ideas.

Our eventual conclusion was that much of the threatened shoreline within the harbor would have to be diked and that constant pumping would be required. During the decades of adaptation it would be important to restrict shoreline development. Possibly major development would be limited to open space or to temporary structures. This would not necessarily preclude redevelopment of shipping in New York, but plans for piers, docks and associated structures would have to address sea-level rise.

Beyond New York Harbor and metropolitan airports, with few exceptions, it would be impossible to protect the entire coastline of Long Island and New Jersey. Land use planners would be forced to anticipate a shifting pattern of development. Coastal pipelines, transportation and utility corridors would need to be protected and relocated. Efforts would be required to foster the migration of wetlands.

WATER SUPPLY

The scenario specifies that precipitation would increase in winter and decrease in summer. Such a shift in precipitation would require more storage capacity in reservoirs due to spring floods and August droughts. An obvious solution is to increase reservoir capacity, but this appears difficult because development pressure already threatens upstate watersheds and reservoirs. Supplying surface water at the same time that sea-level rise increases saltwater intrusion in ground water and in the lower Delaware River Basin, which competes with the same upstate watersheds that serve New York City, will present a major problem. Water conservation will thus become vital throughout the metropolitan area. Conservation measures like gray-water recycling, hardly common today, will be necessary. Ultimately, desalinization may be required.

ENERGY SUPPLY AND DEMAND

Increasing temperatures can be expected to increase demand for air conditioning, which, in turn, increases the demand for electricity. Increased pumping will also increase electrical demand. Such increased demands will come at a time when the regional generation system is already under great pressure because of air pollution controls mandated to reduce atmospheric concentrations of sulfur oxides and ozone. Actions to reduce carbon dioxide emissions will put even more pressure on oil-fired generators.

With demand for electricity increased, but constraints on its production increased as well, imagine all the power plants in the metropolitan area subject to more frequent flooding by storm waters, for most are located on the coast or on rivers! Transmission and distribution lines will suffer storm damage and possible flooding. Interruptions of electrical service will become more common. Emergency back-up sources of power for individual buildings, building complexes and neighborhoods must therefore become the norm.

Large investments will be necessary to protect the electrical supply system. Higher operating and maintenance costs can be expected. Similar to water supplies, the major action to moderate the problem is conservation. Extensive energy conservation must be implemented. This might be mandated as part of the mitigation process that reduces emissions of carbon dioxide, but, as impacts of global warming increase, increased conservation becomes even more essential. Building designs that take advantage of natural air flow to avoid the need for air conditioning will become more important. Building codes must be in place and rigidly enforced. Continued emphasis on solar and wind technology seems obvious. There is much room for improvement, and research and development efforts by the New York State Energy Research and Development Administration (NYSERDA) and similar organizations will be critical.

PUBLIC SERVANTS

How will officials react to this scenario? Manhattan Borough President Ruth Messinger told the workshop, "Tell me about global warming in 2020, but remember: this is 1994. So break the news to me gently!" She also told the assemblage that the job of officials becomes a lot easier when tripwires are in place. For example, if the predicted sea-level rise were seen to be a reality at a point in the future, certain codes and regulations could automatically come into force.

DISCUSSION OF THE REPORT

JEANNE FOX: Sam, you were talking about ZEVs (zero-emission vehicles) in New York City and trucking out to Long Island. Did you talk at all about trucks in the city? Eliminating them, having them come only at certain hours, having more rail transportation? The goods movement in here, and what it does to transportation in the city?

SAMUEL C. MORRIS III: We really didn't talk about trucks in the city itself, except for trying to reduce the number of trucks going through the city and out the other side. We talked about car traffic in the city and decided that what we needed to focus on primarily was the car traffic coming into the city rather than within the city itself.

MARTIN H. GARRELL: One of the people started us thinking about the old Long Island Rail Road yards that are down in Brooklyn on the other side of the Battery. A corridor like that could be restored as a truck terminal. You would also be using new ferry terminals, and you would rethink your waterfront in terms of the old style: bringing goods into Brooklyn by water. These things need to be looked at again because some of that land is still there.

RAYMOND R. RUGGIERI: I was curious about your assumption that everyone would be impacted. Did you have any discussion of an alternative view? What I mean is whether it is the city, or the region, or the country, or the world. My point is: are we talking about some competitive issue here? Or is literally everybody in the same figurative boat?

MORRIS: I don't think everyone is in the same boat. We know that climate change is not going to happen uniformly around the world. Different regions of the world are going to be affected differently, and that effect is going to depend in part upon how the climate changes where they are and how those changes match with the other resources that they have. I don't think there is any way at this time to predict what the relative advantage of one region over another is going to be. The New York area, to the extent that we are in competition with other major financial centers of the world, most of them are in a similar condition. They are on the coast. Hong Kong is probably worse off than New York in this regard.

PHILIP JESSUP: Early in your presentation, you said that because climate change effects would be quite severe that Federal and state action on the mitigation side would probably preclude anything happening locally. We'd have mandates coming down from the Federal government, and so forth. But later you went on to talk about the variety of things which would be undertaken within local government. Were you proposing that even though there would be Federal mandates there is a role for local government? Or should mitigation be left entirely to the Feds and the states, and the cities and local and regional governments should just concentrate on . . .

MORRIS: What we eventually decided—it wasn't a consensus, but it was sort of the view that was adopted—was that the mitigation part . . . We couldn't spend our time doing both. So we decided that the time we spent would focus more on adaptation than mitigation. That we would recognize that the city was going to be required to a certain amount of mitigation. We're not saying that the city or the state shouldn't take any initiatives of their own in that regard. It was just that we as a group couldn't do everything.

Fox: Following up on that. This is a very philosophical question. Everybody is constantly hearing about Federal mandate, Federal pay, state mandate, state pay. The impacts though will be to a large degree local. So that who has the responsibility of paying for that, and dealing with that? And should even the Federal government, or even the state government require certain mitigation efforts, because I think this Federal-state mandate pay thing is going to stay here for a long time. We are going to have resource problems in all levels of government. Philosophically speaking, maybe New York City should decide for itself what they're going to do and how they're going to do it. On the other hand, that's not where I personally come from. (I don't know where my administration comes from on it.) On the other hand, the industrial areas that in these areas that will be first and most impacted are the ones who built up the country. That's where a lot of the population still lies. In fact, this is the financial center of probably the world—second to Hong Kong. Maybe the rest of the country does owe this area something because so much has come out of it and is still coming out of it. So it is a real philosophical thing, and a big picture thing that we are going to have to deal with, in the sense of Washington, Albany, but also locally. But how are we going to deal with it? I think it is something that the academics in the room might want to start doing something about and thinking about more, although I know some of them have.

JESSUP: I think it is also a very practical question because I think that many people have no confidence that the Federal government with lobbies like the auto lobby preventing the passage of new CAFE standards. I think a lot of people feel that it may be impossible for Washington to do a lot of the mitigation work that needs to be done, so therefore it is really going to be local government that will have to take the leadership in the absence of mandates or unmandated voluntary programs at the Federal level.

GARRELL: We had a lot of trouble with that one. We talked about it for a couple of days. The big picture, I think, is that global warming, if you believe it's here, as many climatologists do, if you believe it is inevitable, is a planetary problem. Because you don't have a planetary government that is set up to deal with it—we are not at that stage in our civilization yet—then all we've got is nation-states. Nation-states are strapped for funding. Yet, obviously, these problems are too big, if they really come at us, we have to deal with them. The problem is too big to be dealt with by the region alone. There has to be support at all levels, and I think the public has to get behind the problem the way, for example, it has responded to recycling and waste management. There is a public awareness that you can bring into this. Probably it is going to take a while. It's going to take persistence. That's the real headache. But the financing and the responsibility and the regulations have to come from all levels.

MORRIS: As far as mitigation goes—mitigation is reducing greenhouse gas emissions—and what we want to do is reduce greenhouse gas emissions—to the extent that it's necessary—in the cheapest way possible. And therefore we need to reduce them in the places where we can do that in the cheapest way possible. And if those places happen to be in Ohio, then that's where we should be doing it. Now, that doesn't mean necessarily that Ohio should pay for it. There should be an equal sharing of how we pay for it just because the factories in the Midwest may have spawned the emissions doesn't mean that the financial centers of New York didn't benefit equally from that industrial growth. So the payment has to be shared, but the actual mitigation measures should be done where they can most cost-effectively be done. Now, realistically, everyone probably has to do some, just to show that everyone is sharing in the burden, you might say, more than just paying. But I think that if everyone were to have to mitigate to the same degree and it's going to cost a lost more in New York than it's going to cost in Ohio, it's much better to do it where it's cheapest. Now, adaptation is a little further out, and may end up being as costly or more costly than some of the mitigation measures. And, while it may seem to us initially that a lot of the adaptation involves putting dikes around Manhattan or something like that in the coastal areas, we may find that the adaptation effects are in the Midwest also. The change in agriculture or flooding or water supply problems could equally occur there. It is hard to tell in advance. That I think that has to be dealt with as it comes up.

Comments by Review Panel

ANTONIA BRYSON

*Deputy Commissioner, New York City Department of
Environmental Protection
59-17 Junction Boulevard
Corona, New York 11368-5107*

JEANNE FOX

*Regional Administrator
U.S. Environmental Protection Agency, Region 2
26 Federal Plaza
New York, New York 10278*

PHILIP JESSUP

*International Council for Local Environmental Initiatives
City Hall, East Tower, 8th Floor
Toronto, Ontario, Canada M5H 2N2*

RITA MEYNINGER

*Regional Director
Federal Emergency Management Agency
26 Federal Plaza, Room 1337
New York, New York 10278*

EDWARD A. PARSON

*Department of Public Policy
John F. Kennedy School of Government, Harvard University
79 JFK Street
Cambridge, Massachusetts 02138*

RAYMOND R. RUGGIERI

*Director, New York Metropolitan Transportation Council
One World Trade Center, Suite 82E
New York, New York 10048*

DOUGLAS HILL: It is time now to give our panel its turn. I'd like to give you the chance to speak about this program from your own point of view evaluating these scenarios. Were they adequate, believable? Did they leave anything out? And, particularly, what conclusions would you draw from these scenarios? Are there some robust measures that emerge from these scenarios across the board? What is it we should all take home with us?

BRYSON: Well, I think the first impression that you take away is that there is a tremendous challenge here, and I was interested to see that, in fact, to some extent some of the biggest challenges were raised in the no-change scenario. From my perspective, a lot of the planning challenges and changes in the way we do business are needed just to cope with current issues. And when in the second and third scenarios when there were real crisis problems that were presented, the solutions tend to veer much more toward adapting specifically to those crises, which is the way we mostly do business now. And I think what the first scenario really illustrated to us is if we could sit back and deal with our problems in some long-range strategic way we would find that we need serious solutions even to the existing conditions. So I guess what all that leads to is, I think from my perspective as well as from some of the issues that were brought up in the no-change scenario, one of the most important conclusions we can draw from this is that we need to change the way that we do our planning, both from an economic development perspective and an environmental perspective. And I think that those of us who work in the area would like to see more coordination between both of those goals that everybody believes are valid goals. And that rather than identifying environmental problems with specific initiatives that we as a society or a region want to undertake, what we need to do is to look at a holistic sense of—kind of what you have done here already—a set of environmental conditions that we want to promote and a set of economic development goals that we want to promote and make sure that those work together. So that if you are going to develop a recycling industry on an offshore island you have the zoning and whatever other environmental set of conditions in place already that will promote that at the same time it promotes your environmental conditions. We're not at that stage right now in this region. If we reform in that area it might enable us more easily to solve a lot of the other specific problems.

The second sort of broad theme that I would want to see us go forward on is: how do we get society to pay attention to this? How do we get everybody out there to pay attention to this? How do we enhance what I think every scenario mentioned—the education and the consumer-driven aspect of this? And I don't have any answers to that but I would like to point out that in addition to educating everybody about the CO_2 impacts, global warming impacts that we are going to have to deal with under any of these scenarios, there are many other environmental issues that are already part of our structure, part of the mandates that we have to deal with that we are having difficulty educating the public and getting the consensus to do. So I think that the second thing we have to do is identify the linkages between everything that is out there that needs to be done. And when multiple goals coalesce in a single initiative, take advantage of that. From the air quality perspective, when we have to reduce the inefficient way that we travel around—to enable personal mobility but still to reduce the amount of emissions that we pour into the air—we are all going to have to emphasize energy-efficient ways of moving around, which is not something that we do terribly well right now. That will promote multiple goals, not just the existing Clean Air Act goals but the CO_2 reduction goals as well. So we have to identify those linkages and spearhead initiatives around specific measures that can engage the public such as the recycling that was mentioned here, and that's always the example I give as well because for those of us in the field the mystery is—the positive mystery—is how that managed to capture the public attention and how do we replicate that with the other things that we need to do? That's the second broad conclusion that I would take away from this afternoon's presentation.

And I can go on with some of the specifics, but maybe I will yield my time to the next speaker, and if we have anymore time we can talk specifically.

HILL: Thank you. That's very considerate of you, Antonia. Ray, a lot has been said about transportation planning here.

RUGGIERI: Yes, yes. I was very pleased with that. I share Antonia's view of the panel. I saw a lot of ideas in the first one. I was intrigued by the offshore thing. It reminded me of a proposal somebody had made to close down Kennedy airport and perhaps La Guardia and to build an island out of excavation for a

Long Island Sound river tunnel and maybe even a deepwater port. Imagine what an intermodal facility that would be. I thought the ideas were very intriguing there. I liked the second group where they sort of changed the rules because when I went through the assumptions, that's what I wanted to do, too. And I was amazed at the third one who seemed to say "What, me worry? We can adapt." But I understand the point, and I am being a little flippant there.

Coming out with some of the themes that I heard, I am reminded that when I was in college I had to take a physics course, and the only one I could find was Einstein's Theory on Relativity. It's a long story, but the first day of class the professor asked "What do you think relativity means?" I said, "Everything is related." I am still not sure if that was the right answer then, but that's what I was hearing here. The point was transportation, air quality, energy . . . I would add land use and a few other things. Again, I think as Antonia said, the trick is to see where there are common goals, but also it is important to look where there are conflicting ones. We sometimes overlook that.

But I think conflicts are important, too, because we are really talking about trying to get to consensus, and it is important to lay out all the issues and to make some decisions from them. And that gets to the other theme about education. Sometimes I think that "education" is not the right term, because it sometimes can be taken in a pejorative sense, and I don't use it. I use "dialogue." It needs to be back and forth with interest groups, voters, and citizens, and so-called experts or interested parties to really come to an understanding of where there are shared goals and whether there are legitimate conflicts that need to be dealt with. I kept on wondering to myself, until I guess the very end of the presentation, I kept thinking about "Aren't we sort of looking at this as how we use this whole exercise to avoid what we are talking about?" And I was thinking at the end there it's sort of like an ounce of prevention is worth a gallon of cure. Nobody laughed. Anyway, the point is that . . .

HILL: You should have said a fifth.

RUGGIERI: You're right, and I think I am going to hit that. But anyway, I think the real trick here, and the point was made by the woman back here, that using the scenario planning I think is very instructive, and I think we do do it a lot, we just don't then do anything with it. And the point is, where this is relevant is to really sort of lay out what may happen and what impacts it may cause and that maybe some of the things that you could do to avoid it might be preferable. And I will close with a remark that I always try to think of is that we never make the time or take the time or whatever it is to do things right, but we always find the time or money to fix our mistakes. And I guess go back to the ounce of prevention, and using this to make that point. The last thing is, I keep trying to figure out how a subway under water is going to work, and how you get to it.

BRYSON: A submarine.

RUGGIERI: Yes, maybe. You just have to get out of it.

BRYSON: A new form of mass transit.

HILL: Jeanne?

FOX: I am going to make a couple of points that I think are significant, and I ask you to cut me off after five minutes. The EPA in the United States has been working obviously on this type of thing for a while now, and this administration has made global warming and everything that falls from that a priority. The Climate Action Plan was in, and the United States just submitted to the UN this past month, actually October, our submission to the UN—updating that after a year of looking at it. We stated that we are not going to meet our targets of reaching the levels of 1990, and that we have to do more. And I think that probably most people in this room, and hopefully most people in Congress, and most people who know about this issue believe that. Education and dialogue, however, are necessary, and these types of forums are going to have to continue and expand. Even if, though, we are wrong on the four feet of water in 80 years or 90 years, we still know that something significant will be happening, and it will be more than the four inches. So we have to do something and we have to plan and we have to think and we have to talk and then we have to do, and we are going to have to "do" sooner than later. And we are not going to have to do forty years from now; we are going to have to do sooner than that. So clearly, we have to do two things. One of them is lessen the emissions, and plan on lessening the emissions and continue to do that and working together on a regional basis like the Metropolitan Transportation Council. But more than that, secondly, we have to do the mitigation, plan for the mitigation, talk about the mitigation, and start doing the mitigation before it's an after-the-fact and FEMA has to handle it. I guess that one of our goals should be that FEMA wouldn't have to handle any of this.

MEYNINGER: Yes, we'd like to be out of business.

FOX: That should be one of the goals.

MEYNINGER: Seriously. I mean my staff would get a heart attack if I said that.

FOX: They'll be retired by then.

MEYNINGER: But the point is to get the planning done, the states involved, and the localities involved so they can handle this and not come to the federal government. That's the hope of ours, too.

FOX: And as The Little Green Apples said, to me water supply is one of my most important concerns as it is theirs; it is all interconnected. So that what happens to the New York City reservoirs and what happens to the Hudson River, what happens to the Delaware River, with saltwater intrusion down by Philadelphia and Camden, are related and interconnected. For instance, bigger reservoirs would in fact help in the Delaware River because of that saltwater intrusion—additional reservoirs are a possibility. There is discussion about expanding the F. E. Walter reservoir in Pennsylvania to help with the saltwater intrusion problem. That problem is in our face now with a drought situation. Camden and Philadelphia are having saltwater intrusion in their wells now; Cape May County has saltwater intrusion into their wells—that's happening now. So even if there is no other effect than the four inches, we need to deal with that, and that type of linkage is important. Now there is a Delaware River Basin Commission, which includes New York, New Jersey, Pennsylvania, and Delaware. It's a very well-run organization, and it looks at this. It has a lot of very good engineers who look at this, but that has to be expanded to include more impacted governments and people and citizens in general.

Thirdly, regarding infrastructure, there needs again to be some kind of regional coordination, so what group should do that? There is at least the possibility of the Port Authority, which includes two of the states, could take the lead on the infrastructure, the roads and everything in addition to the roads that are considered infrastructure. Somebody has to take the lead in coordinating this type of effort. I think that discussions have to be held sooner than later as to who those somebodies should be to plan for the mitigation part concerning the infrastructure.

Regarding energy, clearly that is one of the big, important things, and again there has to be dialogue and there has to be planning, and I think that is ongoing. Brookhaven has a market allocation model that I am told is pretty good. It looks at the energy future and it looks at the CO_2 emissions and it costs it out. That type of thing has to move out into the world of decision-makers. I just found out about that recently, and I have been involved in this business for a little while. It is something that is discussed and worked with more than the modelers, more than the people who are the hands-on energy people; it has to get to decision-makers in all sectors so that we can start planning for that in the reduction of emissions.

Finally, the discussion that Sam Morris was having about local benefits versus who should pay, who should control; I think it's clearer to me that it's a local benefit to a large degree. That will be primarily a local benefit for us along the coastline because we are avoiding flooding and everything else. On the other hand, I think it is necessary to have national regulation. We can't do it locally. It is necessary because, as he said, you really need to do it at the cheapest cost. And that means it's got to be national, it can't be local. In addition, there is the fairness issue that I discussed. For instance, I have a chart here, TABLE 3, Demand-side Opportunities for Reducing CO_2 (Morris *et al.*) From this it is very clear that a lot of these things in the chart are much cheaper than some other things that are being discussed. For instance, mass transit—while I am a strong supporter of it—costs a lot more to reduce CO_2 than something else, for instance, heavy duty trucks, fuel efficiencies, and working on that. Again, it's got to be more than just a regional thing.

So it is something that we need to philosophically think about and bring to the attention of politicians and those types of decision-makers: that in fact the local benefit will have to come from, to a large degree, a national effort. And so when they say federal-mandate, federal-pay, the federal government is saying you need to do this. In fact, the benefit is to the local area, because they are the ones who will not be inundated because we are lessening these emissions.

Bottom line: we need more dialogue like this, and more education of decision-makers and stakeholders, so they can start thinking about the philosophy of this problem, and start getting their minds into it. So that later on when things start happening they understand it, and just don't react as this country too often does.

HILL: Thank you. Bill, with your experience in other cities dealing with the CO_2 emission problem . . . ?

JESSUP: What I would like to do is propose a framework drawing on some of the results of the scenarios that might help you on your way to sell these results to elected officials. What I do in my work, I try to enlist cities around the world. We have quite a few now that are developing local action plans to reduce greenhouse gas emissions. But they are doing that kind of work in relation to other things that need to be done in their jurisdictions. What we do is try to develop ways to help decision-makers create those links and set priorities.

One way to think of the task at hand is using the diagram with three circles (FIG. 1). This circle representing the powers of local government—the Port Authority of New York and New Jersey, the City of New York, the boroughs, other governments around—that's represented in this circle. In this other circle are the kind of adaption strategies that have come out of your individual scenarios. And over here are prevention strategies. These are the things that over the long term, as all of us around the globe undertake these measures, will be able to prevent global warming. If we were just to do adaptation, we'd be looking at this area, this being the area where local government has jurisdiction and can do something about a particular infrastructure problem. Same thing here, in terms of prevention. Local government has control over urban growth management; national government does not. So local government has a lot of power to determine what kind of human settlements we have; the federal government doesn't. So where these three circles intersect, it seems to me, is where you're going to get the most

FIGURE 1

synergy among the different strategies, both on the adaptation side and on the prevention side.

An example is water conservation. It is not surprising actually that water conservation came out in all three scenarios. Whether we have no climate change at all, or whether we have an extreme case of it, everybody thought it was a good idea. And I think the reason is that it is something that makes sense in its own right, and something that we need to do to plan for the future.

Then what we should try to do to better integrate the adaptation and the prevention sides is to adopt a kind of cost framework or matrix where we can evaluate the adaptation strategies, the prevention strategies, and then see what the net social cost is to society around that impact. So if we took something like water supply, the costs of adaptation, of expanding the reservoir system, building desalinization plants, and so forth, these would be negative. These are costs. Water conservation is going to pay over the long term because it is going to create revenues. We balance these off, we may actually have a plus side here and the benefits we get over here can help offset the costs from this side. This is particularly true in terms of energy conservation because you're creating revenue streams—money—from energy savings that could actually go into a pool to help pay for some of the adaptation costs.

So I think that if you can—this is just one suggestion—but if you can find some way to integrate strategies on both sides and then link that with the powers of local governments, you come up with a list of measures that are really sellable.

RUGGIERI: Call his group *Two Bites of the Apple*.

HILL: And finally, we have Ted Parson, a scholar of global climate policy in the international arena.

PARSON: And an enthusiast for exercises like this. I am going to talk briefly about three or four of the different reasons that you would undertake an exercise like this, and then talk about some of the results of the different groups in the contexts of those purposes. Why would you spend two days trying to do scenario planning about global climate change rather than trying to think about it in other more conventional ways?

The first question you might ask is: what are other more conventional ways of thinking about it? Well, we can do planning by sort of taking last year's numbers and incrementing them by one percent per year, or we can try to use some kind of formal models of the economy or of the environment or of the global climate change system. There are obvious limitations to both of those, when you are thinking about long-term, big, imperfectly understood issues like climate change.

I suggest that there are four big reasons that you do this. The first thing is, an exercise like this is guaranteed to knock you out of your incremental thinking, and knock you beyond the items that are on your immediate political agenda by requiring you to think about the problem in a specific way set some distance in the future. So I really commend the organizers for doing something as potentially useful as this.

The second thing it does, it makes you think about it in vivid, specific ways. Ruth Messinger, over lunch, said the purpose of exercises like this is to make policy-makers believe in the possibility of the issue you are talking about. I really want to emphasize that. I think it is one of the biggest values of it. But to do that, it really takes a lot of hard work. I mean, you have to somehow exercise your narrative imagination to tell a coherent, detailed, vivid story of different possible futures. And then subject that to all kinds of critical testing, because it's really hard to know—I mean, we can't know anything—it is hard even to have reasonable, plausible predictions about what the world will look like in 2020 or 2070. But there are obvious things that you can do if you work hard, if you work in groups that bring together knowledge across a bunch of disciplines as you have tried to do today, to look in detail at some of the detailed implications of the things you are assuming, test them for consistency, make yourself try out specific things.

That's why I was really impressed by some of the groups who had some specific things in the future, like we've talked a lot about comprehensive coastal redevelopment plans for New York City. Group 3 had this wonderful idea of diking the three entrances to the harbor and pumping the Hudson. Group 2 had fishing and aquaculture development in New York Harbor waters. What I think is the wonderful potential of exercises like this, and what I would urge you to do more of, is take one of those things and say, OK, we are going to dike the harbor and pump the Hudson. Where will we build it? Where will we get the land? Whose authority was invoked in order to do it? What were the political difficulties involved in getting them to exercise that authority? How much did it cost? How do we raise the money? Now, how's the coastline look in the harbor after we've done it? Hmm. Who was using it before for what purposes? Are they still using it for the same purposes? Are there things, are there major uses of the coastline that have become infeasible? Are there things that have become attractive and feasible that weren't before? Those people who are using it now: what are they doing now instead of? And the people who aren't using it anymore, what are they doing instead of what they were doing?

There are systematic ways that you can discipline your thinking to go down these tracks, not to make specific predictions but to exercise your imaginations and exercise your sense of coherence and consistency. I think it is a wonderful, and it's a vastly underused exercise for those reasons. Vehicles for testing consistency that you should use: physics, chemistry, engineering, budgets. If you propose a solution that involves a certain cost, of the order of tens of billions, then think carefully about where that money is coming from. How the taxes were raised, and what it is not being spent on instead.

An interesting example of an exercise like this that I'd just like to inform you about: the Swedish Economic Planning Council a few years ago tried to organize a comprehensive nationwide long-term environmental planning exercise. They used a wonderful device to try to push these directions of vividness and consistency. They got a bunch of eminent national artists to draw paintings of the landscape of Sweden in the year 2030 under different specific assumptions about the energy system and the forms of economic development and industrial development and the kind of environmental policies and controls that were in place. They used those paintings as centerpieces to focus the thinking and discussion of scenario-planning exercises.

That's actually a very interesting exercise. It is a great antidote for certain forms of simplistic ideology and certain forms of popular wisdom, for example, the picture of the landscape under the all-renewables system. Well, a lot of the environmentalists found that they really didn't like that picture very much. I don't know how many of you have seen the windpower farms in California, but anybody who can look at those and assert that massive development of certain kinds of renewables is an utterly environmentally benign form of energy development is, in my view, dreaming. The discipline of visual portrayal of the implications of different decisions really helped in this Swedish planning exercise.

Last, perhaps the most important thing that an exercise like this can do is identify those things that you most need to learn, that you don't yet know. Because in disciplining yourself to follow these paths you say, OK, we're going to do this. What does that mean? Who's building it? Who's using it? What did the construction phase look like? Where did the money come from? You find quickly as you follow these branched trees of contingencies that you come to points where there are potentially knowable pieces of information that you don't know, or that nobody knows yet, but we could because we haven't spent the money or the resources trying to find out.

Group 2, I think, highlighted this when their chair or rapporteur said, "We realized part way through our work we really don't know enough about linkages, about particular certain kinds of linkages." In an exercise like this, one of the most useful things you can do is actually be specific about that: what linkages do you not know enough about? What ones look like they are the most important for your understanding of potential futures? And articulate how you can go about finding those things out.

There are certain kinds of criticism of exercises such as this that really aren't particularly useful. I'd say the criticism, "Did you consider X?" is not particularly useful, because on everybody's list of important things to think about there will be a long, a large number of X's that it is simply not possible to think about. Really, the discipline of this exercise has to do with moving you into the future in a way that your imagination of contingencies and causal relations shapes what's important, and you necessarily have to exclude all kinds of other important things.

OK. Those are my first-cut ideas about why you would do such a thing, and it has some implications about how you would evaluate the outcome of such an exercise.

In terms of doing such an exercise for global change, what are the questions that you would most hope to answer out of such an exercise? I suggest that the first and most important one is: how important is global climate change? We can't necessarily presume that for a particular jurisdiction like New York City that it's necessarily going to be that big a deal. That's why I think it's a great idea, for one thing, to include the no-change and the medium estimate groups, rather than just have extreme groups. We're talking about 25 years in the future. What are the things that are going to be the most salient problems and the most salient causes of the changes in people's lives over that time?

Maybe it's going to be global climate change. That's part of the purpose of an exercise like this. Maybe it's not. What's going to happen to technological change, immigration and emigration, change in the structure of the local economy? That question is going to be different whether you are looking at New York City, the region, the United States, the world, or specific other places.

Second, you can ask: what are the most significant other paths of effect?

I was struck by the extent to which most groups concentrated most of the time on sea level rise. If you had divine insight by which you got to know what the most important pathways of effect of global climate change on life in New York City was in the year 2030 or 2040, I'd be willing to bet a lot of money that sea level rise in New York City will not be the most important. It may be the one that you can see most clearly and identify most clearly, and I think that's the reason that everybody likes to concentrate on it. It's so visual, right? You know, up a foot, what's that mean? Draw the contour lines, build the dikes, how much does it cost? I wouldn't be at all surprised if the most important impacts on New York City actually came from indirect climate effects elsewhere in the world that affect the economy and the population here through immigration, emigration, change in economic structure, and so on.

Two more points. In doing this for a small region it's hard but important to think coherently about the relationship between abatement and adaptation. New York is a little place in the context of this issue. By my order-of-magnitude calculations, in the scenario we are talking about, New York City will be 0.2 percent more or less of the world's population, and of the order of one percent of world emissions. So you can't assume a coupling between the rigor of efforts at emissions abatement made here and the magnitude of impacts observed here. It's tempting. I didn't hear anybody explicitly make this error, but I heard a sort of a mixing of the two sides of the thing throughout a lot of the discussion, as if you might get tempted into thinking that somehow rigor of effort at abating emissions will somehow be rewarded by lesser impacts? Uh-uh. No, I'd actually suggest that it's useful to think of two dimensions of scenarios if such an exercise is repeated. Combine high and low rigor of efforts locally to abate emissions with high and low impacts. And think in each of those four cells. Because the political linkages that you can imagine will be very, very interesting. You might imagine, for example, that extreme local effects, whether or not they are directly attributable to climate change—I mean whether you can legitimately do it—might create kind of local political motivation and consensus for doing more. Or you might imagine that if things come sort of late and slow, that we're just shocked into a kind of apathy and inaction.

Last points, timing and probability. First, it would be useful to do such scenario planning exercises with explicit consideration of the time dimension, because the same magnitude of change could be very different in the magnitude and character of difficulties it posed depending on whether it came gradually or suddenly. This would suggest a format for exercises in which, say, three separate rounds are conducted representing the years 2000, 2020, and 2040, and the team's job is to tell an evolving future history under the information revealed in each round, always being locked into the consequences of past decisions they had already made.

A related suggestion, which I describe as "probabilities," is that since a primary objective of planning for long-run environmental change is identifying robust policy strategies, it would be useful for teams to start out their planning exercises not knowing which of several possible scenarios they were operating under, and letting realizations from prior probability distributions be made as the exercise proceeds. You could even imagine running scenario time back and forth, examining the implications of the same early policy decisions under several future realizations of the intensity of change.

Glossary

AC	Alternating current
A/C	Air conditioning
APWA	American Public Works Association
AVO	Average vehicle occupancy
Btu	British thermal unit
C	Carbon
CA	California
CAAA	Clear Air Act Amendments of 1990
CAFE	Corporate average fuel economy (U.S. passenger automobile standard)
CBD	Central Business District
CFC	Chlorofluorocarbon
CFL	Compact fluorescent light
cm	centimeter
CN	Connecticut
CNG	Compressed natural gas
CO	Carbon monoxide
CO$_2$	Carbon dioxide
Con Edison	Consolidated Edison Company of New York, Inc.
DC	District of Columbia; direct current
DCP	Department of City Planning
DOS	Department of State
DOT	Department of Transportation
DSM	Demand-side management
EEM	Energy efficiency measures
EIS	Environmental Impact Statement
EMS	Energy management system
EPA	U.S. Environmental Protection Agency
FCCC	United Nations Framework Convention on Climate Change
FDR	Franklin D. Roosevelt
FEMA	Federal Emergency Management Agency
FL	Florida
ft	foot
GPF	gallon per flush
GPM	gallon per minute
GSP	Gross State Product
HOV	High-occupancy vehicle
HVAC	Heating, ventilation and air conditioning
ICLEI	International Council for Local Environmental Initiatives
IPCC	Intergovernmental Panel on Climate Change
IRT	IRT Division of the New York City Transit Authority (subway)
ISTEA	Intermodal Surface Transportation Act of 1991
ITS	Intelligent transportation systems
IVHS	Intelligent vehicle highway systems
JFK	John F. Kennedy
km	kilometer
kWh	kilowatt-hour
KY	Kentucky

LILCO	Long Island Lighting Company
LIRR	Long Island Rail Road
m	meter
MARKAL	A widely used energy system model
MBtu	Million Btu
MESA	Marine EcoSystems Analysis
MI	Minnesota
mm	millimeter
mph	miles per hour
MSRC	Marine Sciences Research Center, SUNY at Stony Brook
MSW	Municipal solid waste
MTA	Metropolitan Transportation Authority
MW	megawatts
N	Nitrogen
NASA	National Aeronautics and Space Administration
NC	North Carolina
ND	North Dakota
NFIP	National Flood Insurance Program
NGVD	National Geodetic Vertical Datum of 1928
NJ	New Jersey
NOAA	National Oceanic and Atmospheric Agency
NO$_x$	Nitrogen oxides
NPS	National Park Service
NY	New York
NYC	New York City
NYPA	New York Power Authority
NYS	New York State
NYMARKAL	Computer model of New York State energy system
NYSEO	New York State Energy Office
NYSERDA	New York State Energy Research and Development Authority
O	Oxygen
OR	Oregon
O$_3$	Ozone
OH	Hydroxyl radical
OTA	U.S. Office of Technology Assessment
PA	Pennsylvania; The Port Authority of New York and New Jersey
PAN	Peroxyacetyl nitrate
PANYNJ	The Port Authority of New York and New Jersey
PATH	Port Authority Trans-Hudson Corporation (subway)
PM10	Particulate matter of 10 microns or less
ppb	parts per billion
ROI	Return on investment
RPA	Regional Plan Association
SCR	Silicon rectifier
SUNY	State University of New York
T$_e$	Effective radiating temperature
TSP	Total suspended particulates
TV	Television
UEC	Unit energy consumption
UN	United Nations
USACE	U.S. Army Corps of Engineers
USGS	U.S. Geological Survey

GLOSSARY

VA	Virginia
VAV	Variable air volume
VMT	Vehicle-miles traveled
VOC	Volatile organic compounds
VSD	Variable speed drive
ZEV	Zero-emission vehicle
°C	Celsius temperature
°F	Fahrenheit temperature
°K	Kelvin temperature
µg/m³	microgram per cubic meter

Subject Index[a]

Air pollution. *See also* Emissions
 aerosol particles and, 92
 climate change and, 91–101
 health effects of, 3–4, 91–92, 101–107
  mitigation of, 107–108
 lead and, 92
 policy goals on, 4
 pollutants. *See also* Air quality
  EPA-regulated, 92
 possible impacts, 3–4
Air quality. *See also* Air pollution; Emissions; Greenhouse gases
 air conditioning and, 14
 global climate change and, 97–98
 meteorological factors affecting, 96–97
 observations on, 98–100
 ozone problem, 93–94
  climate change and, 97–100
  health effects, 104–107
  models, 100–102
  NO_x and, 3–4, 92
  photochemistry, 94–96
  temperature and, 92–93, 97–101
 planting to improve, 52
American Society of Civil Engineers, 157

Baked Apple conference. *See also* Global warming; Greenhouse effect
 conclusions, 12–16
 recommendations, 16–17
 summary, 1–14
Baked Apple conference, public policy remarks, 157–159
Bight Restoration Program, 88

Carbon dioxide (CO_2). *See under* Emissions
Carbon monoxide (CO). *See under* Emissions
Chicago Urban Forest Ecosystem Project, 51
Chlorofluorocarbons (CFCs). *See under* Emissions
Clean Air Act Amendments of 1990 (CAAA)
 and emission compliance, 4
 and infrastructure planning, 3
 State Implementation Plans for, 80
 and traffic reduction, 13, 123
 and trip reduction program, 8–9, 118

Climate change. *See also* Global warming
 air conditioning and, 14, 146–147
 and air quality, 91–101
 Climate Change Action Plan, 1
 and energy supply and demand, 140–141
 health effects of, 101–109
 local consequences of, 15
 projected, 1–2
Coastal erosion, 5–7. *See also* Waterfront planning
 Arverne case, 48–49
 Fair Harbor case, 48
Coastal Zone Management Act, 44, 87
Connecticut
 defense procurement reductions and, 30–31
 employment forecast for, 33
 employment growth distribution in, 36
 payroll employment changes in, 32
 population growth characteristics in, 40
 waterfront planning in. *See* Waterfront planning

Demand-side management (DSM), 12

East River Water Quality Plan, 87
Emissions. *See also* Air pollutants; Clean Air Act Amendments; Greenhouse gases
 carbon dioxide (CO_2)
  from cars/trucks, 8, 19, 111, 158
  factors affecting amount of, 113–114
  fossil fuels and, 9–10
  and greenhouse effect, 19, 21, 97
  impacts of, 19
  ocean reservoir of, 24
  reduction of, 4, 9, 114–123
  source of problem, 112–113
 carbon monoxide (CO), 92, 111
 chlorofluorocarbons (CFCs), 19, 111
 nitrogen oxides (NO_x)
  and acid rain, 111
  auto traffic and, 8
  and ozone formation, 3–4, 92, 94–96, 111
  reduction of, 4, 151–156
  transportation and, 8, 19, 111, 158
  reduction of, 76–78
 volatile organic compounds (VOC)
  and acid rain, 3–4, 111
  auto traffic and, 8

---

[a] Page numbers in italics indicate remarks made in discussion.

and ozone formation, 3-4, 92, 94-96, 111
Energy. *See also* Energy efficiency measures
  in buildings
    end uses for, 125
    residential UEC, 74
    systems, 125-126
  conservation of
    benefits of, 12
    codes, 9
    DSM and, 10-12
    and transportation emission reduction, 76-78
  global climate change and, 139
  greenhouse effect and, 10-11
  infrastructure use of, 78-79
  New York State codes, 9
  NYMARKAL model, 11-12
  supply and demand, 139
    climate change effects on, 140-141
    policy trends affecting, 141-142
    uncertainties in planning, 142-143
Energy efficiency measures (EEMs)
  impacts of, 126-128
  implementation of
    barriers to, 129-131
    incentives for, 129
  and minimizing emissions, 129
  moving toward solutions, 131-133
  in New York City, 9-10
  in New York State, 11
  other options, 128
  unit energy consumption and, 74
  upgrades, 128
Environmental impact statement (EIS), 49
Environmental Law Section (NYS Bar Association), 157
Environmental Protection Agency (EPA)
  air pollutants regulated by, 92
  Green Lights program, *137-138*

Federal Emergency Management Agency (FEMA)
  educational material of, 17
  flood hazard regulations, 7, 46, 49
  Hurricane Evacuation Computer Model, 56
  and sea-level rise formula, 53
Flood control. *See also* Waterfront planning
  FEMA flood hazard regulations and, 46
  greenways agenda and, 7, 49, 52
  and infrastructure damage prevention, 70-72

ISTEA and, 49
need for, 13
Flooding
  and coastal erosion, 5-7
    Arverne case, 48-49
    Coastal Zone Management Act, 44, 87
    Fair Harbor case, 48
    National Flood Insurance Program, 80
    prevention of. *See also* Flood control
    greenways and, 7, 49, 52
    structures for, 54, 70-72
    zoning for flood hazards, 53
  sea-level rise and, 2, 45
  storm of 1993, 48, 50-51
Framework Convention for Climate Change (FCCC), 1, 12

Global warming. *See also* Climate change; Greenhouse effect; Heat
  conditions associated with, 58-59
  consequences of, 1-2
  and flood control, 49-51
  greenhouse gases and, 19 25
  health effects of, 2-4, 102-107
    mitigation of, 107-108
  and infrastructure. *See also* Land use policy/management/planning recommendations, 79-80
    scenario 1: response, 61-72
    scenario 2: response, 72-79
    vulnerability of, 57-59
  IPCC Scientific Assessment estimates, 21-23
  and O_3 levels, 3-4
  policy goals on, 4
  and sea-level rise
    adaptation to, 52-54
    IPCC estimates of, 58
    and waterfront planning, 45-48
  and unemployment, 14
  and waterfront planning, 43-54
  and water resources, 85-88
Greenhouse effect. *See also* Air pollution; Air quality; Climate change; Global warming
  climate models of, 20-21
    key results from, 21-23
    uncertainty sources, 23-25
  conference on
    conclusions, 12-16
    planning context, 2-3
    recommendations of, 16-17
  consequences of, 2
  critical uncertainties future, 41
  energy concerns, 2
    efficiency/conservation, 9-10

SUBJECT INDEX

and energy system, 10-11
  NYMARKAL results, 11-12
flooding/coastal erosion concerns
  mitigation/adaptation measures, 6-7
  possible impacts, 5-6
health effects
  of climate change, 101-104
  health risks, 3-4
  mitigation options, 107-109
physics of, 19-20
projected climate change, 1-2
radiatively active gases and, 19-20
sustainable economic future and, 40-42
  Long Island, 39-40
  New York City, 39
and transportation planning, 7-9
United Nations and, 1
and water supply, 4-5
water supply concerns, 4-5
Greenhouse gases. *See also* Air pollution; Air quality; Emissions
climate responses to, 21-23
New York State's ability to reduce, 143-146
radiatively active gases, 19-20
reduction strategies, 144-148
and transportation planning, 7-8
Greenways
and flooding prevention, 7, 52
ISTEA and, 49

Infrastructure. *See also* Global warming; Land use
damage prevention
  flood control structures, 70-72
  material/structural design modification, 68
  support facilities relocation/redesign, 68-69
  water dependency reduction, 69-70
global warming consequences for, 58-59
land use interrelationships with
  and global warming response/retardation, 72-79
  sea-level rise and, 60-72
as major energy user, 78-79
recommendations for, 79-80
vulnerable facilities, 62
  airports/ports, 65-66
  bridges/roadways, 63-64
  damage prevention requirements for, 68-72
  energy production/transmission facilities, 67
  for hazardous materials, 67
  transit tunnels, 65
  for transportation, 63-66
  for water pollution control, 66
water dependency of, 61-62
Institute for Marine and Coastal Sciences (Rutgers U.), 157
Intergovernmental Panel on Climate Change (IPCC), 1, 3
  Scientific Assessment, 21-23
  sea-level rise estimates of, 58
Intermodal Surface Transportation Efficiency Act of 1991 (ISTEA)
  and bicycle networks, 118
  and greenways agenda, 49
  and infrastructure planning, 3, 80
  and land use planning, 80
  and transportation planning, 9, 13, 123

Jamaica Bay Comprehensive Watershed Plan, 87

Land use
distribution of, 75
and global warming retardation, 61, 72-79
infrastructure interdependency with, 60-61
planning processes for, 80
recommendations for, 79-80
sea-level rise and, 60-72
Long Island
defense procurement reductions and, 30-31
development scenario for, 39-40
employment forecast for, 33
employment growth distribution in, 35
payroll employment changes in, 32
Long Island Sound Study, 87

Marine EcoSystems Analysis (MESA), 58
Marine Sciences Research Center (S.U.N.Y., Stony Brook), 157
Metropolitan New York
demography/settlement pattern, 36
  population age changes, 39
  population growth, 37
  by race/ethnicity, 38
  regional growth characteristics, 37-40
development scenarios
  critical uncertainties future, 41
  sustainable economic future, 40-41
economic/demographic outlook for, 29-30, 41-42

employment forecast, 33
employment growth distribution, 34–36
employment growth outlook, 31–34
national context, 30–31
payroll employment changes, 32
recent economic performance, 31
greenhouse effect and
air quality, 91–101
health effects, 101–109
Metropolitan Transportation Authority (MTA), 69
Municipal Leaders' Summit on Climate Change and the Urban Environment, 1

National Bridge Rehabilitation Program, 68
National Environmental Policy Act, 44
National Estuary Program for New York-New Jersey Harbor, 88
National Flood Insurance Program, 80
National Geodetic Vertical Datum (NGVD), 64–65
National Park Service (NPS), natural hazard policy of, 7, 46
New Jersey
employment forecast for, 33
employment growth distribution in, 36
payroll employment changes in, 32
population growth characteristics in, 40
waterfront planning in. *See* Waterfront planning
New York Academy of Sciences, 157
New York Bar Association, 79
New York City
DEP planning studies, 87
development scenario for, 39
employment forecast for, 33
employment growth distribution in, 34–35
energy efficiency of, 9–12
flood control system, 70–71
flood effects on, 6
infrastructure of
global warming effects on, 58–59
water dependency of, 61–62
land uses distribution, 75
payroll employment changes in, 32
traffic
and CO_2 emission, 7–8
congestion in, 8–9, 13
and VMT reduction, 8–9
vulnerable transit facilities, 64–65
water conservation measures, 12–13
waterfront of. *See also* Waterfront planning
and infrastructure, 61–62
water supply of
climate change consequences to, 2
conservation of, 5
and potential conflicts, 13–14
sources of, 4–5
New York State
Bridge Inventory and Inspection System, 63
and energy conservation
DSM program, 10–12
energy codes, 9
energy system
capacity of, 10–11
efficiency of, 11
NYMARKAL model of, 11–12, 139
and greenhouse gas emissions reduction, 143–146
home rule decisions in, 3
Hurricane Evacuation Computer Model for, 53
New York State Energy Office, *149*
New York State Energy Plan, 79
New York State Greenhouse Study, *149*
Nitrogen oxides (NO_x). *See under* Emissions
NYMARKAL model
for air conditioning CO_2 emissions, 147–148
and energy supply-and-demand, 139
of New York State energy system, 11–12

Ozone (O_3). *See under* Air quality

Port Authority of New York and New Jersey, 157
and climate change response, 16
PATH system, 69
Public Service Commission (New York State), 12

Recycling program, *160–162*
Regional Plan Association (RPA), 157
and climate change response, 16
and waterfront revival, 49

Scenario planning reports
assuming accelerated climate change
discussion of, *197–199*

SUBJECT INDEX

and energy supply/demand, 196
mitigation/adaptation, 194
public servants' reaction, 196–197
the scenario, 193–194
and transportation, 194–195
waterfront effects, 195
and water supply, 196
assuming medium climate change
air quality/health effects, 186
building design/maintenance, 186
demographic/economic forecasts, 183–185
discussion of report, *187–191*
energy demand/supply, 186–187
FEMA comments, *187–189*
land use/infrastructure planning, 185
other recommendations, 187
transportation planning, 186
waterfront planning, 185
water resources/wastewater management, 185–186
assuming no climate change, 177–182
discussion of report, *181–182*
climate assumptions for, 164
introduction to, 163–175
review panel comments, *201–210*
Sea Grant Institute, 157
Sea-level rise
adaptations to, 52–54
believability of, 45
FEMA formula, 53
and flood duration, 70
and infrastructure/land use, 57–59
levels expected, 6
MESA studies of, 58
and National Flood Insurance Program, 80
and vulnerable infrastructure, 62–67
waterfront planning and, 43, 45, 48–49, 51–52
and water supply, 85–87
Statewide and Substate Strategies (New York State), 87

Traffic. *See also* Transportation planning
and CO_2 emissions, 7–8
reduction, 114
congestion reduction, 13, 119–122
and greenhouse gas emission reduction, 119–123
and VMT reduction, 8–9, 114–119
Transportation planning. *See also*
Intermodal Surface Transportation Efficiency Act
and energy conservation/emission reduction, 76–78

for greenhouse effect, 7–9, 111
and greenhouse gas emissions. *See also*
Air pollutants
CO_2, 112–114
daily VMT reduction, 114–121
historical context, 111–112
reduction, 119–123

Unit energy consumption (UEC), 74
U.S. Army Corps of Engineers (USACE)
flood hazard policy of, 7
inventory of transportation systems, 80
transit facility vulnerable points, 64–65
transportation infrastructure vulnerability, 63–67
U.S. Geological Survey (USGS), 67
U.S. Office of Technology Assessment (OTA)
global warming study of, 78
greenhouse gas emissions attribution, 73, 77

Vehicle-miles of travel (VMT). *See also*
Clean Air Act Amendments of 1990;
Transportation planning
and CO_2 emissions
amount of, 112–113
factors affecting, 113
and fuel emission factor, 114
reducing growth of, 8–9, 114–119
taxis and, *124*
and traffic congestion, 8, 119–122
Volatile organic compounds. *See under*
Emissions

Water conservation
desalinization and, *89*
measures for, 12–13
motivation for, 135–136
moving toward solutions, 131–136
options for, 134–135
waste reduction limitations, 135
Waterfront planning
and development
rethinking, 75–76
zoning, 51
economic development/political factors, 45–48
EIS and, 49
environment and, 44–45
and flood control, 5–7, 49–51
zoning for, 53
and global warming prevention, 52
greenways agenda, 49

private beach renourishment, 48
proposed plan for Manhattan, 158
regulatory requirements stiffening, 48–49
renaturalizing bulkheaded shorelines, 49–52
sea-level rise and, 52–54
waterfront opportunities, 43–44
waterfront reuse marketing, 47–48
Water pollution
  Clean Water Act and, 87
  control facilities, 66
  Water Pollution Control Act, 44
Water Resources Council (New York State), 87
Water supply
  conservation, 12–13
  decision-making locations, 87–88
  ecosystem protection approach, 88
  estuarine system policy, 87
  mitigation/adaptation measures, 5
  policy for, 85–87
  possible greenhouse effect on, 5
  sources, 4–5
  systems at risk, 85

Index of Contributors

Alpern, R., 85–90
Armstrong, R. B., 29–42
Audin, L., 125–138

Bowman, M. J., 163–175
Broccoli, A. J., 19–27
Bryson, A., 201–210

Chin, P. A., 183–191

Falcocchio, J. C., 111–124
Fox, J., 201–210

Garrell, M. H., 193–199
Goldstein, G. A., 139–150

Haff, H. B., 43–56
Hill, D., xi–xii, 139–150

Jessup, P., 201–210

Kleinman, L. I., 91–110

Lipfert, F. W., 91–110

Meyninger, R., 201–210
Messinger, R. W., 157–162
Mooney, A., 177–182
Morris, S. C. III, 139–150, 193–199

Parson, E. A., 201–210

Reaven, 177–182
Ruggieri, R. R., 201–210

Sanghi, A., 139–150
Sullivan, E. O., 151–156
Swanson, R. L., 177–182

Tanikawa-Oglesby, S., 183–191
Tusa, W., 183–191

Zimmerman, R., 57–83